Acknowledgements

This report is the result of a research project carried out by Buro Happold in partnership with Bath University under contract to CIRIA. The project was developed and managed by Ann Alderson, senior research manager at CIRIA, with advice and guidance from a project steering group, whose support and valuable contributions are gratefully acknlowledged. The steering group comprised:

Mr Martin J M Wilson (chairman)	Taylor Woodrow Construction Ltd
Mr David Beardsworth	Mechtool Engineering Ltd
Mr Gordon Bishop	NetComposites (formerly with PERA)
Mr John Cadei	Maunsell Ltd
Mr Richard Cole/Mr Bruce Ellis	formerly Permali UK Ltd
Mr Steve Cole	Concargo Ltd (for first part of project)
Mr Anders Edgren	Reichhold UK Ltd
Dr Sue Halliwell	Building Research Establishment Ltd
Mr Martin Starkey	SP Systems Ltd
Dr Les Norwood/Mr David Taylor	Scott Bader Co Ltd
Mr Peter Jennings	ACO Technologies plc
Ms Sarah Kaethner	Ove Arup & Partners
Mr Marcus Simmons	DERA
Mr Michael Stacey	Brookes Stacey Randall
Mr Mark Thompson	Sheppard Robson
Dr Peter Thornburrow	Vetrotex International
Mr Peter Woodward	WS Atkins Ltd (representing the DETR)

Corresponding members

Mr Neil Burford	University of Dundee
Mr Angelo Manesero	Dutco Balfour Beatty
Dr Geoff Turvey	Lancaster University

The project was funded by:

Department of the Environment Transport and the Regions through the Partners in Innovation scheme

CIRIA core members	Concargo Ltd
ACO Technologies plc	Permali UK Ltd
Mechtool Engineering Ltd	Scott Bader Co Ltd
Reichold UK Ltd	Vetrotex International
SP Systems Ltd	Buro Happold

CIRIA and the author[s] gratefully acknowledge the support of these funding organisations and the technical help and advice provided by the members of the steering group. Contributions do not imply that individual funders necessarily endorse all views expressed in published outputs.

Principal authors:

Andrew Cripps	(Buro Happold)
Bryan Harris	(University of Bath)
Tim Ibell	(University of Bath)

Contributing authors from Buro Happold:

Andrew Best (*structural design*)	Ollie Kelly (*structural design*)
Ian Liddell (*structural design and procurement*)	James Maclean (*product design*)
Adam Monaghan (*fire*)	Michael Taverner (*structural design*)
Syreeta Robinson Gayle (*environment/materials*)	Dominique Pool (*architectural/ case studies*)
Karim Yngstrom (*civil/bridge engineering*)	

Note

Recent Government reorganisation has meant that DETR responsibilities have been moved variously to the Department of Trade and Industry (DTI), the Department for the Environment, Food and Rural Affairs (DEFRA), and the Department for Transport, Local Government and the Regions (DTLR). References made to the DETR in this publication should be read in this context.

For clarification, readers should contact the Department of Trade and Industry.

Fibre-reinforced polymer composites in construction

Cripps A

Supported by

Harris B

Ibell T

Fibre-reinforced polymer composites in construction

Cripps A

Construction Industry Research and Information Association

CIRIA C564   © CIRIA 2002   ISBN 0 86017 564 2

Keywords		
materials, materials technology, design and buildability, building technology		
Reader interest	**Classification**	
clients, engineers, architects	AVAILABILITY CONTENT STATUS USER	Unrestricted Guidance document Committee guided clients, engineers, architects

Contents

List of tables

List of figures

Glossary and abbreviations

ABBREVIATIONS

ABS	Acrylonitrile Butadiene Styrene copolymer
BMC	Bulk Moulding Compound
CFRP	Carbon-Fibre-Reinforced Polymer
CFM	Continuous Filament Mat
CSM	Chopped Strand Mat
DMC	Dough-Moulding Compound (known in the USA as bulk-moulding compound, BMC)
FRP	Fibre-Reinforced Polymer
GRP	Glass-Fibre-Reinforced Polymer (NB. GFRP is not used in the UK to avoid confusion with American usage to imply Graphite-Fibre-Reinforced Polymer
ILSS	Interlaminar Shear Strength
MDF	Medium Density Fibreboard
NDI (NDE)	Non-Destructive Inspection (Evaluation)
NDT	Non-Destructive Testing
PAN	Polyacrylonitrile
RIM	Resin-infusion moulding (also formerly used to signify resin-injection moulding)
RTM	Resin-Transfer Moulding
VARIM	Vacuum-Assisted RIM
VARTM	Vacuum-Assisted RTM
SMC	Sheet Moulding Compound

GLOSSARY

accelerator: Also known as a promoter, a material mixed with a catalyzed resin to speed up the chemical reaction between the catalyst and resin so that the reaction can take place at room temperature rather by the application of heat. Common accelerators for polyesters are salts, like cobalt naphthanate and tertiary amines, and are often supplied premixed in the resin.

ageing: The effect on materials of exposure to an environment; the process of exposing materials to an environment for an interval of time.

anisotropic: Exhibiting different properties in response to mechanical or thermal stresses applied along axes in different directions.

aramid: Aromatic polyamide fibres characterised by good high-temperature, flame-resistance, and electrical properties. Aramid fibres are used to achieve high strength, high modulus reinforcement in polymer composites. Sometimes referred to as poly-aramid (trade names Kevlar or Twaron).

aspect ratio: The ratio of length to diameter of a reinforcing fibre.

autoclave: A closed pressure vessel used for curing laminates under pressure and heat.

autoclave moulding: After lay-up, the entire assembly is placed in an autoclave. The additional pressure achieves higher fibre-to-resin ratios and improved removal of air.

axial winding: In filament-wound reinforced polymers, a winding with the filaments parallel to the axis.

bag moulding: A technique in which the consolidation of the material in the mould is affected by the application of fluid pressure through a flexible membrane.

barcol hardness: A hardness value obtained by measuring the resistance to penetration of a sharp steel point under a spring load. The instrument, the Barcol Impressor, gives a direct reading on a scale of 0 to 100. The hardness value is often used as a measure of the degree of cure of a polymer.

bi-directional laminate: A reinforced polymer laminate with the fibres oriented in various directions in the plane of the laminate; a cross-plied laminate.

blister: Undesirable rounded elevation of the surface of a polymer, whose boundaries may be more or less sharply defined, resembling a blister on the human skin. The blister may burst and become flattened.

bond strength: The amount of adhesion between bonded surfaces; a measure of the stress required to separate a layer of material from the base to which it is bonded.

bulk moulding compound (BMC): (see dough moulding compound)

burst strength: Hydraulic pressure required to burst a vessel of given thickness; commonly used in testing filament-wound composite structures.

catalyst: A substance that changes the rate of a chemical reaction without itself undergoing permanent change in its composition; a substance that markedly speeds up the cure of a compound when added in small quantity compared with the amounts of primary reactants. Catalysts for polyester resins are usually organic peroxides. Catalysts are chemically unstable, and are usually supplied in combination with other more inert substances.

chopped-strand mat (CSM): A form of reinforcement consisting of sheets of randomly arranged chopped fibres some 5 cm in length held lightly together with a binder which may or may not be soluble in matrix resins.

cohesion: (1) The propensity of a single substance to adhere to itself. (2) The internal attraction of molecular particles toward each other. (3) The force holding a single substance together.

composite: A homogeneous material created by synthetic assembly of two or more materials (a selected filler or reinforcing elements and compatible matrix binder) to obtain specific characteristics and properties. Composites are subdivided into the following classes on the basis of the form of the structural constituents; *fibrous*: the dispersed phase consists of fibres; *flake*: the dispersed phase consists of flat flakes; *laminar*: composed of layers of laminate constituents; *particulate*: dispersed phase consists of small particles; *skeletal*: composed of a continuous skeletal matrix filled by a second material.

compression moulding: A technique for moulding thermoset polymers in which a part is shaped by placing the fibre and resin into an open mould cavity, closing the mould, and applying heat and pressure until the material has cured or achieved its final form.

compression strength: (1) The ability of a material to resist a force that tends to crush. (2) The crushing load at the failure of a specimen divided by the original sectional area of the specimen.

compressive stress: The compressive load per unit area of original cross section carried by the specimen during the compression test.

contact moulding: A process for moulding reinforced polymers in which reinforcement and resin are placed on an open mould, cure is at room temperature using a catalyst-promoter system or by heat in an oven, and no additional pressure is used.

continuous filament: An individual flexible fibre of glass of small diameter and great or indefinite length; made into continuous-filament mat (CFM)

continuous-filament yarn: Yarn formed by twisting two or more continuous filaments into a single continuous strand.

continuous roving: Parallel filaments coated with sizing, gathered together into single or multiple strands, and wound into a cylindrical package. It may be used to provide continuous reinforcement in woven roving, filament winding, pultrusion, prepregs or high strength moulding compounds, or it may be used chopped.

core: (1) The central member of a sandwich construction to which the faces of the sandwich are attached. (2) A channel in a mould for circulation of heat-transfer media.

coupling agent: Any chemical substance designed to react with both the reinforcement and matrix phases of a composite material to form or promote a stronger bond at the interface; a bonding link.

creel: A device for holding the required number of fibre bobbins in the desired position for unwinding.

creep: The change in dimension of a polymer under load over a period of time not including the initial instantaneous elastic deformation; at room temperature it is called cold flow.

cure: To change the properties of a resin by chemical reaction, which may be condensation or addition; usually accomplished by the action of heat or catalyst, or both, and with or without pressure.

curing agent: Also called a Hardener, a reactive agent added to a resin to cause polymerization. Curing agents participate in the polymerization process, for example of two-component epoxy resins, and become part of the final molecular network of a thermosetting resin. The ratio of resin to hardener is critical for complete cure. Hardeners may be latent, in which case the resins which contain them are curable only at elevated temperatures, or they may be activated at room temperature, often by the addition of an initiator or accelerator (*qv.*).

cycle: The complete, repeating sequence of operations in a process or part of a process. In moulding, the cycle time is the elapsed time between a certain point in one cycle and the same point in the next.

dough moulding compound (DMC): A viscous dough compounded from a pre-catalyzed thermosetting resin, a filler powder (such as chalk), short glass fibres (1 cm to 2 cm long), and other (often proprietary) processing and moulding aids. The resin cures when the dough is squeezed under pressure to the required finished shape in a heated mould.

drape: The ability of pre-impregnated woven cloth to conform to an irregular shape; textile conformity

E glass: A borosilicate glass; the type most used for glass fibres for reinforced polymers; suitable for electrical laminates because of its high resistivity. (Also called electric glass).

elasticity: The property of materials by virtue of which they tend to recover their original size and shape after deformation (as distinct from plasticity).

elastic limit: The greatest stress that a material is capable of sustaining, without permanent strain remaining upon the complete release of the stress. A material is said to have passed its elastic limit when the load is sufficient to initiate plastic or non-recoverable deformation.

epoxy resins: Polymers based on resins made by the reaction of epoxides or oxiranes with other materials such as amines, alcohols, phenols, carboxylic acids, acid anhydrides and unsaturated compounds.

exotherm: The liberation or evolution of heat during the curing of a polymer product.

fabric: A material constructed of interlaced yarns, fibres, or filaments, usually planar.

fatigue: The failure of a material or decay of its mechanical properties after repeated applications of stress. (Fatigue tests give information on the ability of a material to resist the development of cracks, which eventually bring about failure as a result of many repeated cycles).

fibre orientation: Fibre alignment in a non-woven or a mat laminate where the majority of fibres are in the same direction, resulting in a higher strength in that direction.

filament winding: A process for fabricating a composite structure in which continuous reinforcements (filament, wire, yarn, tape, or other) impregnated with a matrix material, either previously or during the winding, are placed over a rotating removable form or mandrel in a prescribed way, to meet certain stress conditions. When the required number of layers has been applied, the wound form is cured and the mandrel removed.

filler: A relatively inert material added to a plastic mixture to reduce cost, modify mechanical properties, serve as a base for colour effects, or improve the surface texture.

flame-retarded resin: A resin that is compounded with certain chemicals to reduce or eliminate its tendency to burn.

flexural strength: (1) The resistance of a material to breakage by bending stresses. (2) The strength of a material in bending expressed as the tensile stress in the outermost fibres of a bent test sample at the instant of failure. For most materials this value is usually higher than the normal tensile strength. (3) The unit resistance to the maximum load before failure by bending.

fracture energy: The energy required to break a sample of material containing a sharp notch in a test often carried out under impact conditions; a measure of the toughness or impact resistance of a material.

gel: The initial jelly-like solid phase that develops during the formation of a cured resin from a liquid.

gel point: The stage at which a liquid begins to exhibit pseudo-elastic properties. (This stage may be conveniently observed from the inflection point on a viscosity-time plot). Also referred to as gel time.

gel coat: A filled resin applied to the surface of a mould before lay-up. The gel coat becomes an integral part of the finished laminate and is usually used to improve the surface qualities of the material (eg the environmental resistance).

glass transition: The reversible change in an amorphous polymer or in amorphous regions of a partially crystalline polymer from (or to) a viscous or rubbery condition to (or from) a hard and relatively brittle state. The glass transition generally occurs over a relatively narrow temperature region and for processing purposes is analogous to the solidification of a liquid to a glassy state although it is not a phase transition. The glass transition temperature, T_g, is the approximate midpoint of the temperature range over which the glass transition takes place.

hardener: (see Curing Agent) A substance or mixture added to a polymer composition to promote or control the curing action by taking part in it.

impact strength: The ability of a material to withstand shock loading; the work done in fracturing a test specimen in a specified manner under shock loading.

impregnation: In reinforced polymers, the saturation of the reinforcement by the matrix resin.

injection moulding: A technique developed for processing thermoplastics, which are heated and forced under pressure into closed moulds and cooled.

inhibitor: A substance which retards a chemical reaction; used in certain types of monomers and resins to prolong storage life.

interface: The junction surface between two different media; the contact area between the reinforcement and the laminating resin.

interlaminar shear strength (ILSS): The shear stress to cause shear failure or delamination between layers of a laminated material.

laminate: A composite material consisting of one or more layers of fibre, impregnated with a resin system and cured, sometimes with heat and pressure.

laminate ply: One layer of a product, which is itself produced by two or more layers of materials.

lay-up: (1) As used in reinforced polymers, the reinforcing material placed in position in the mould. (2) The process of placing the reinforcing material in position in the mould. (3) The resin-impregnated reinforcement. (4) The component materials, geometry etc, of a laminate.

mandrel: (1) An internal former for hollow components. (2) In filament winding, the core around which resin-impregnated fibre is wound to form pipes, tubes, or vessels.

matched-die moulding: A reinforced-polymer manufacturing process in which matching male and female metal dies are used (similar to compression moulding) to form the part.

matrix: In reinforced polymers the matrix is the resin binder, which consolidates the fibre bundles and converts them to an engineering solid.

modulus of elasticity: The ratio of the stress or applied load to the strain or deformation produced in a material that is elastically deformed. Also called Young's modulus when the stress/strain relationship is linear.

monocoque: A form of structure, like a fuselage or motorcar body, in which all structural loads are carried by the skin. Such a structure is often capable of being formed in a single piece and by a single operation from composite materials. In a semi-monocoque, loads are shared between a skin and a framework, which provides local reinforcement for openings, mountings, etc.

monomer: A simple molecule that is capable of reacting with like or unlike molecules to form a polymer; a monomer or mer is the smallest repeating structure of a polymer; for addition polymers, the monomer represents the original unpolymerized compound.

mould: The cavity in or on which the moulding material is placed and from which it takes form.

nesting: In reinforced polymers placing plies of fabric so that the yarns of one ply lie in the valleys between the yarns of the adjacent ply (nested cloth).

NDI (Non-Destructive Inspection): A process or procedure for determining material or part characteristics without permanently altering the test subject. Non-destructive testing (NDT) and non-destructive evaluation are roughly synonymous with NDI.

platens: The mounting plates of a press, to which the entire mould assembly is bolted.

polymer: A high-molecular-weight organic compound, natural or synthetic, whose structure can be represented by a repeated small unit (mer), eg polyethylene, rubber, cellulose. Synthetic polymers are formed by addition or condensation polymerisation of monomers.

precursor: For carbon fibres, the rayon, polyacrylonitrile (PAN), or pitch fibres from which carbon fibres are made.

polyamides: Polymers in which the structural units are linked by amide or thioamide groupings. Many polyamides are fibre-forming.

polyesters: Thermosetting resins produced by dissolving unsaturated alkyd resins in a vinyl-type active monomer, such as styrene, methyl styrene, and diallyl phthalate. Cure is effected through vinyl polymerization using peroxide catalysts and promoters, or heat, to accelerate the reaction. The resins are usually furnished in solution form, but powdered solids are also available.

post-cure: Additional elevated temperature cure, usually without pressure, to improve final properties and/or complete the cure. In certain resins, complete cure and ultimate mechanical properties are attained only by exposure of the cured resin to higher temperatures than those of curing.

post-forming: Processing of flat sheets of reinforced thermoplastics by re-heating above the glass transition temperature and pressing or stamping to a given final shape. Reinforced thermosets cannot be post-formed.

preform: A pre-shaped fibrous reinforcement of mat or cloth formed to the desired shape on a mandrel or mock-up before being placed in a press mould or RTM tool.

pre-impregnation: The practice of mixing resin and reinforcement and effecting partial cure (B-staging) before use or shipment to the user (see also Prepreg).

prepreg: Ready-to-mould material in rolled-sheet form, which may be cloth, mat or fibres impregnated with resin and stored for use. The resin is partially cured to a soft and slightly sticky B stage and supplied to the fabricator, who lays up the finished shape and completes the cure with heat and pressure.

pressure-bag moulding: A process for moulding reinforced polymers, in which a tailored flexible bag is placed over the contact lay-up on the mould, sealed, and clamped in place. Fluid pressure, usually compressed air, is exerted on the bag, and the part is cured.

pultrusion: A process for the manufacture of rods, tubes, and structural shapes of a constant cross section. After passing through a resin dip tank, bundles of rovings are drawn through a heated die to form the desired cross section.

resin: A solid, semi-solid, or pseudo-solid organic material, which has an indefinite (often high) molecular weight, exhibits a tendency to flow when subjected to stress and usually has a softening or melting range. Most resins are polymers. In reinforced polymers the material used to bind together the reinforcement material, the matrix (see also Polymer).

resin-transfer moulding (RTM): A moulding process in which catalysed resin is transferred into an enclosed mould into which the fibre reinforcement pre-form (*qv.*) has been placed; cure may be accomplished with or without external heat. RTM combines relatively low tooling and equipment costs with the ability to mould large structural parts.

roving: The term is used to designate a collection of bundles of continuous filaments either as untwisted strands or as twisted yarns. Rovings may be lightly twisted, but for filament winding they are generally wound as bands or tapes with as little twist as possible. Glass rovings are predominantly used in filament winding.

S glass: A magnesia-alumina-silicate glass, especially designed to provide filaments with very high tensile strength (known as R glass in France).

scrim: A non-woven or open-weave reinforcing fabric made from continuous-filament yarn in an open-mesh construction. Often used to support an adhesive film.

size (or sizing): Any treatment consisting of starch, gelatin, oil, wax, or other suitable ingredient applied to yarn or fibres at the time of formation, to protect the surface and facilitate handling and fabrication or to control the fibre characteristics. The treatment contains ingredients, which provide surface lubricity and binding action but, unlike a finish, no coupling agent. Before final fabrication into a composite, the size is usually removed by heat-cleaning and a finish is applied.

sheet-moulding compound (SMC): A compound containing similar constituents to DMC (*qv.*) but with longer fibres (5 cm to 10 cm) which is supplied in sheet form for press moulding between heated dies.

shell: A thin-walled structure in which all stresses are in the plane of the shell (ie through-thickness stresses are all zero).

spray-up: Technique in which a spray gun is used as the processing tool. In reinforced polymers for example, fibrous glass and resin can be simultaneously deposited in a mould. In essence, roving is fed through a chopper and ejected into a resin stream, which is directed at the mould by either of two spray systems.

stitching: Placing of fibres normal to a laminated plate (so called z-direction fibres) to improve delamination resistance and interlaminar shear strength.

strain: The resultant change in dimension of a material when subjected to a stress, expressed as a fraction or percentage of the original dimension (it is a dimensionless quantity).

stress: Most commonly defined as engineering stress; the ratio of the applied load (P) to the original cross-sectional area (A).

tape: A form of unidirectional prepreg consisting of continuous fibres that are aligned along the tape axis parallel to each other and held together purely by the impregnating resin.

thermoplastic: A polymeric material capable of being reversibly softened by increase of temperature and hardened by decrease in temperature; applicable to those materials whose change upon heating is substantially physical rather than chemical and which can be shaped by flow into articles by moulding and extrusion. Examples include, polypropylene, ABS, Nylon, polyethylene.

thermoset: A polymeric material, which changes into a substantially infusible and un-mouldable material after it is cured by application of heat or by chemical means. Although a thermoset material will soften at its Tg, it will never return to its pre-cured liquid state. Examples include epoxy, unsaturated polyester and phenolic resins.

thixotropic: Gel-like at rest but fluid when agitated; having high static shear strength and low dynamic shear strength at the same time.

tow: A large bundle of continuous filaments, generally 1000 or more, usually designated by a number followed by "k," indicating multiplication by 1000; for example, 12 k tow contains approximately 12 000 individual filaments.

vacuum-bag moulding: A process for moulding laminates in which a sheet of flexible material is placed over the lay-up on the mould and sealed. A vacuum is applied between the sheet and the lay-up. The entrapped air is pulled out of the lay-up and removed by the vacuum. Atmospheric pressure provides the consolidation pressure.

viscosity: The property of resistance to flow exhibited within the body of a liquid or semi-solid expressed in terms of relationship between applied shearing stress and resulting rate of strain in shear.

warp: (1) The yarn running lengthwise in a woven fabric; a group of yarns in long lengths and approximately parallel, put on beams or warp reels for further textile processing, including weaving. (2) A change in dimensions of a cured laminate from its original moulded shape.

weave: The particular manner in which a fabric is formed by interlacing yarns. In a plain weave, the warp and weft fibres alternate to make both fabric faces identical; in a satin weave, the pattern produces a satin appearance, with the warp tow over several weft tows and under the next one (for example, eight-harness satin would have warp tow over seven weft tows and under the eighth).

weft: The transverse threads of fibres in a woven fabric running perpendicular to the warp; also called fill, and woof.

wet-out: The condition of an impregnated roving or yarn wherein substantially all voids between the sized strands and filaments are filled with resin.

yarn: An assembly of twisted fibres or strands, natural or manufactured, to form a continuous yarn suitable for use in weaving or otherwise interweaving into textile materials.

Executive Summary

INTRODUCTION

Fibre-reinforced polymer composites is a general term used to describe a wide range of products made up of a combination of fibres in a matrix material. These materials are used extensively, particularly in the marine, aerospace and wind turbine industries, where their high strength to weight ratios and good performance in harsh environments mean that they are the best choice. Approximately one million tonnes per year are produced by the European composites industry across all applications, and this increased by approximately 6 per cent in 1999.

In the construction industry fibre-reinforced polymer composites are widely used in applications such as cladding, pipes, for repair and in strengthening work. Construction makes up around 30 per cent of the total market for FRP composites, second only to the automotive sector. However, there are many situations where they are not used. This may be because alternative and better understood materials are able to meet the requirements of the project, for significantly lower initial costs. There are other conditions however, where the best solution would be the use of FRP composites, but they are still not being used.

This report seeks to address the reasons why FRP composites are not used more widely in construction, and to encourage their appropriate use in the future. This is carried out through a series of steps that aim:

- to increase the confidence (through case studies) of designers and clients in the ability of the materials to "deliver the goods" over the desired life time
- to demonstrate the types of application that are practical now or could be in the future
- to help designers to choose the "right" materials and design with them.

WHAT IS AN FRP COMPOSITE?

In general engineering terms, a composite is a combination of two or more materials used together for any reason. This work is only concerned with fibre-reinforced polymer composites – the abbreviation FRP is used in this report. These consist of a "fibre" in a polymer-based "matrix".

The fibres carry the main loads, and it is because of their excellent strength-to-weight properties that FRP composites have successful applications. These are surrounded by a "matrix", which serves to bind the fibres together, transfer loads between them, and protect them from the effect of the outside environment and physical impact. It is the combination of the two materials that enables the FRP composite to be beneficial.

WHY USE COMPOSITES?

There are a wide range of reasons why an FRP composite may be the best material for a particular job and these are addressed through this report. FRP composites can offer added value through their high strength-to-weight ratio as compared to other materials, their ability to survive harsh environments, and the fact that they can be formed into complex shapes. Further they can be fire resistant, and their low weight brings installation benefits in space-cramped and time-critical projects. The combination of these factors means that the composite solution can be cheaper than any other alternative, particularly in terms of whole-life cost. This is in spite of the relatively high cost per kilogram of the raw materials.

It is vital to emphasise that the beneficial use of FRP composites usually follows from designing to use its strengths and not simply to mimic some other material. The item developed is usually best thought of as a complete product, and not a set of components as is generally the case with other materials.

SCOPE OF THIS REPORT

The scope of this report, and the project work that supported it, is the application of fibre-reinforced polymer composites in the construction industry. It addresses many potential applications, attempting to balance the wide variety of possibilities with the need to provide more detail in key areas.

The construction industry is taken to include both building and civil engineering sectors, but excludes industrial plant. The use of unbonded fibres as reinforcements for concrete will not be covered, but the use of FRP composite rebars and stressing tendons will be. The large amount of work on the repair/ strengthening of bridges with composites is recognised, but not repeated in depth.

Glass fibres have the largest share of the market for the fibres in the composites, but this report considers all types of synthetic fibres. Natural fibres are used in some circumstances, but these are not discussed further here.

HOW TO USE THIS REPORT

This report is structured with the aim of helping the reader through the process of understanding FRP composites, in order to gain the confidence to use them.

The first chapter introduces the materials, discusses the overall strengths they have and how weaknesses can be overcome. It introduces the themes that will be developed through the rest of the report.

Chapter 2 gives an outline of the manufacturing processes used to make FRP composites. There are a variety of techniques used, and the language used to describe them can be confusing at first. However it is useful to understand how the techniques differ, and the potential for each one to produce different products. This section is also important for helping to understand sales and other literature from the industry.

In order to reinforce the way in which these materials are used, Chapter 3 gives a set of case studies showing examples of what has already been built with FRP composites. These examples explain why others have chosen to use them, and the effectiveness of the solutions developed. Where possible there is information on the current state of the buildings or structures, to support the view that FRP composites have a long service life.

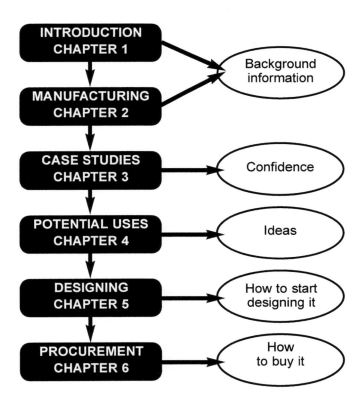

Figure 0.1: *Flow diagram for the report*

Chapter 4 considers what is possible with FRP composites. This is the result of discussions with people inside the industry, and from outside, and has suggestions for areas that look likely to be suited to future use of FRP composites. In this respect it is vital to work with the strengths of the materials to find the right applications. The aim of this part is to inspire the reader to develop their own ideas, using the thought processes described here, and to enable them to discuss these ideas with FRP composite specialists.

Having understood what has already been done (Chapter 3) and what can be done in the future (Chapter 4), Chapter 5 addresses how to go about designing with the materials. This starts from how to choose the type of materials to be used, and when to do so in comparison with other "conventional" materials. It then discusses the key design areas: structural, fire performance, joining, finishes, environmental resistance and environmental impact. This chapter contains more technical data than the others, and is aimed more at designers than clients. However, it is not intended to be comprehensive design guidance, but to give an indication for the direction to take and where to seek further advice.

Chapter 6 is about procurement. Given the range of manufacturing techniques, there are many different companies involved, and general guidance is given on how to get your design produced and completed on site.

This book should enable the reader to feel more confident about using FRP composite materials in the future, and show how to go about doing this in a way that is successful for all concerned.

1 Introduction to fibre-reinforced polymer composites

1.1 WHAT IS A FIBRE-REINFORCED POLYMER COMPOSITE?

A fibre-reinforced polymer (FRP) composite is a combination of fibres within a matrix of a plastic or resin material. The fibres bring the strength to the composite, the matrix binds the fibres together, transfers the loads between them and the rest of the structure, and protects the fibres from the environment. One of the best known FRP composite materials is glass-fibre-reinforced polymer – GRP. This has a large share of the market, but there are many other materials available. The fibres are usually one of the following:

- glass
- carbon
- aramid (often known through the tradenames Kevlar or Twaron)
- natural fibres (used less often, for example hemp)

The fibres can be used in three significantly different ways, with the performance changing for each.

1 The highest performance in terms of strength in one direction comes from **uni-directional** fibres. The fibres are parallel and give their maximum possible performance in this single direction.
2 By arranging the fibres in a weave or mat, strength can be gained in more directions, although the limit strength is reduced.
3 By chopping the fibres into short lengths and arranging them randomly, equal strength is achieved in all directions. This is generally the cheapest technique, used for the least structurally demanding cases.

These issues are discussed further in the section on manufacturing, and in the design section.

The **matrices** are divided into two major groups of polymers: thermosets and thermoplastics.

Thermosets are resins that are cured by heat or chemical reaction to become hard, rigid substances that cannot be softened by reheating. The most commonly encountered thermoset polymers that are used as matrices for fibre-reinforced polymer (FRP) composites are:

- polyesters
- vinyl-esters
- epoxies
- phenolics.

Thermoplastics melt to form highly viscous liquids at elevated temperatures (usually between 200°C and 300°C) but solidify to glassy or crystalline substances as they cool. They can be softened again by reheating. Examples are:

- polyamides (nylon)
- thermoplastic polyesters (eg PET)
- polypropylene

All of these materials can be used on their own, but can benefit from the reinforcing provided by the fibres.

1.2　ADVANTAGES OF FRP COMPOSITES

There are many reasons why people use FRP composite materials, and they vary between applications. FRP composites have been selected on a cost/programme basis due to savings or added value generated by one or more of the issues discussed below. The benefits offered by FRP composites vary depending on the choice of resin, fibre and process of manufacture. There needs to be a process of optimisation of the design of the composite since not all of the best properties can be achieved at the same time. It is vital when considering FRP composites to remember that it is the whole-life cost, rather than the simple cost of the product that is most likely to be the positive feature.

The examples referred to below are all discussed in the cross-referenced case studies in Chapter 4 of the report, and these reflect the different benefits of the materials. The main benefits are usually one or more of the following.

1 Weight saving.

2 Time saving: its high strength-to-weight ratio means that components can be light and so construction time can be reduced in time-critical projects, eg bridge repairs (Section 3.3).

3 Able to add to a structure: its high strength to weight ratio means you can add to structures without further strengthening of existing structure (Section 3.3).

4 Valuable in repair/strengthening: possible to bond to existing structure *in situ* to allow the repair of structures *in situ* following damage, or their strengthening to allow for increased loads (eg repairs in Section 3.3) or developing design requirements.

5 Low maintenance requirements: ideal where access is difficult or expensive, eg for high roofs, underwater (Section 4.2.4).

6 Resists a harsh or corrosive environment eg road de-icing salts: limits need for repair/maintenance for harsh environments eg off shore, sewage systems, chimneys, boats.

7 Impact/blast resistant: can be designed to absorb blast or ballistic loads and other impacts (Section 4.3.1).

8 Fire resistant: although untreated materials do burn, FRP composites can be designed to meet the most stringent fire requirement.

9 Freedom of shape: moulding techniques allow a unique shape/complex geometry eg Millennium Dome Rest Zone (Section 3.5.3).

10 Variety of appearance: Can give a particular colour or texture eg Liverpool Cathedral, architectural applications (Section 3.3.6). The appearance is up to the designer.

11 Radio transparent: Can be transparent to electromagnetic radiation e.g Radomes (Section 3.1.6) Stone Mountain (Section 3.1.4), mobile phone company headquarters (Section 3.1.8), large structural frame (Section 3.1.9).

12 Non-conducting and non-magnetic: brings safety benefits near to high power electrical systems or for mine sweeping boats.

1.3 FACING THE CHALLENGES

There are many challenges to be faced when designing with any material. Some specific issues often raised as barriers to the use of FRP composites are introduced below and discussed in more detail in later sections.

1.3.1 Codes and standards

There are relatively few codes and standards for FRP composites, compared to longer established materials like steel and concrete. This is a discouragement to the use of the material in some sectors, and for this reason considerable effort is being made to develop standards. The main codes at present are limited to pultruded profiles and the construction of vessels and tanks (BSI 1987 and 1999 respectively, see references).

1.3.2 Environmental impact

We have seen a growth in interest in environmental issues in recent years, and it is therefore important to consider these in the context of FRP composites. In a very simple analysis it may appear that they will not score well, because of their oil-based production, the chemicals used in their production, and difficulties with recycling. However in environmental analysis, more than any other area, a holistic approach is needed, and here the materials are much more positive.

FRP composites are strong from an environmental perspective when they are able to "save" an otherwise failing structure, if this would otherwise result in demolition and rebuilding with new materials. They can result in systems with a far greater life span, (eg flue gas chimneys, sewage plant) again reducing impacts. Some FRP composites can limit fire impacts (see Section 1.3.3). They can result in reduced supporting structure, where their weight reduces load, and in the automotive sector in better fuel performance – again because of weight savings.

Most resins are based on by-products from the oil industry. As this is driven by the need to produce liquid fuels for transport and heating, the other products are less critical in terms of resource depletion. In fact they can be considered as useful applications of an otherwise low-value material, and it has been estimated that only 5 per cent of the by-products of petro-chemical manufacture are being used, so there would be a benefit in using more.

Furthermore the energy input in producing FRP composites is low, compared to metals or cement, and in general less FRP composite is needed than other materials. The fact that most plastics are insulators is also helpful, potentially reducing the in-use energy of a building, which would generally exceed the energy used in producing the materials.

This is not to say that FRP composites are an ideal material, from an environmental perspective, and careful design is needed to minimise impacts, as with any material. More detailed discussion is given in Section 5.2.5.

1.3.3 Fire

Most polymers will burn when exposed to fire, like many other materials. Just as with timber for example, this affects the way in which the materials can be used, but does not prevent their use. Similarly, designing with steel requires fire engineering of the structure to minimise risk to occupants and fire fighters in the event of fire, and a related approach is needed for FRP composites.

With the exception of most phenolic resins, all untreated matrices will burn when exposed to flame under the wrong conditions. Careful design and the right choice of resins, additives and fillers however, can make FRP composites fire retardant. In fact some combinations of materials perform better than almost any other material in fire and are used in some fire walls, and extensively on oil drilling platforms, on account of their ablative behaviour. Of course these choices will give a higher first-cost solution than the simplest materials. FRP composites are generally poor thermal conductors, so the heat of a fire does not spread in the way that it does with metals. The reduced quantity of materials needed for equal strength can reduce the amount of combustible material in some circumstances. The smoke from fire can be a concern, and so careful choices need to be made for internal applications.

As with all issues in design there is always a compromise to be made between the different material properties needed for a project. If fire resistance is absolutely critical then FRP composites can often be the most cost-effective solution. This same level of performance would not be needed in other situations, and good fire engineering and general design will be able to achieve the right degree of safety, at a reasonable cost – and meet all of the other design conditions.

These issues and how they can be addressed are discussed in full in Chapter 5.

1.3.4 Joining

As with all other materials, joining is a key technology for FRP composites. In order to benefit from the properties of the FRP components, it must be possible to combine them efficiently with each other and with other parts of the system.

The choice is between adhesive technologies and mechanical fixing, although sometimes a combination (hybrid jointing) is used. These are discussed further in Section 5.2.1. The adhesive technologies are attractive as they use the same basic materials as the resins themselves, but they are hard to de-construct. Mechanical fixings are simple to understand, but bring problems with stress loading. An area that deserves further development, is "snap joint" or "clip together" systems, for very rapid construction.

1.3.5 Perceived design complexity

An FRP composite product can be produced in almost any shape and with a huge range of mechanical properties, able to meet most requirements. It is ironic that the great diversity of potential uses of FRP composites means that many people are put off using them because they seem too difficult. The designer needs to decide not just whether to use an FRP composite, but also which fibre, which resin and how they should be combined. The issues of procurement (discussed later) should also be addressed. This contrasts with the longer established and more standardised nature of design with some other materials, where standard products are available off the shelf.

This report aims to show that the flexibility of design is a great benefit, and the experience of others who have used the materials before shows that the process is not too difficult. There are products, which are to some extent "standard", and advice is available. The industry is working to improve further in both respects.

1.3.6   Finishes/aesthetics

FRP composites have a particular strength over most other materials, because they can emerge from the manufacturing process ready to install onto the structure with no further treatment required. An effectively limitless range of colours can be included into the final surface (gelcoat), and these colours have a long life (see for example the case study on the Hermann Miller factory in Bath, Section 3.1.12). Furthermore the technology of surface gel coats has improved since early examples, which did not work as well as was hoped.

It is also possible to simulate a wide range of textural appearances using FRP composites, including wood and stone effects. The "fake wood" doors do not need the same treatment that wooden ones do, and can result in a lower cost. Some people do not like this type of mimicry, but at times it is essential. Others should be encouraged to use the technology to develop novel aesthetics.

The FRP composites industry is trying to change its image with respect to aesthetics, as one observer put it, "so that people will think more Lotus Elan and less Reliant Robin". Having said that, it would be better still if they learn that both have used FRP composites extensively, and both have been successful in their own way. Finishes and aesthetics are discussed further in Section 5.2.6.

1.3.7   Health and safety

The construction industry has a relatively poor record on health and safety, and is making a continued effort to improve this. FRP composites, like other materials, have an impact on this. Most of the hazards involved with the materials (styrene gas release or dermatitis from chemicals) are controlled in the manufacturing processes, and do not affect site workers. Minimising of site finishing is important in this respect, and proper training in the use of adhesives and how to cut FRP composites is also important.

On the positive side, the low density of FRP composites will generally assist with reducing a different class of hazards relating to the movement of heavy building components, particularly in cramped locations.

Health and safety is discussed further in Section 5.2.4.

1.4   THE CURRENT MARKET FOR FRP COMPOSITES

According to APFE (Association of European Glass Fibre Producers) statistics, building and construction represented 31.3 per cent of the total European composites market in 1997 (including infrastructure), accounting for 129 170 tonnes of reinforcing fibres.

The breakdown within building and construction was given as follows:

- **Roofing** – 20 per cent. Specific applications include: porch roofs, roof trims, gables, domes, spires, roofing sheets and roof lights, canopies, structural beams and roof trusses, rainwater gutters, roof tiles and artificial slates, shingles.

- **Façade cladding** – 6.4 per cent. Specific applications include: cladding panels and components, concrete formwork, aggregate panels.

- **Flooring** – 7.7 per cent. Specific applications include: floor panels, skirting boards, walkway gratings, screed reinforcement, drainage gratings and covers, reinforced PVC floor coverings.

- **Sanitaryware** – 10.7 per cent. Specific applications include: baths, shower trays, sinks and wash basins.

- **Electrical** – 23 per cent. Specific applications include: electrical cabinets, lighting components, electricity and gas meter boxes.

- **Decoration/architectural** – 4.4 per cent. Specific applications include: cornices and architectural features, door skins.

- **Infrastructure** – 18.6 per cent. Specific applications include: power, radio and telecommunications towers, utility poles, rebars for concrete, mining/tunnel rock bolts and support tubes, cooling towers, bridge enclosure panels, bridge decking, telephone booths, structural beams, storage tanks, pipework, column wrapping (seismic protection), bridge strengthening plates, concrete formwork, water tanks, street furniture.

- **Other** – 9.2 per cent. Specific applications include: insulation profiles for glazing, double glazing unit spacers, railings and fence posts, roller shutter doors, door frames, window shutters, conservatory frames, garage doors.

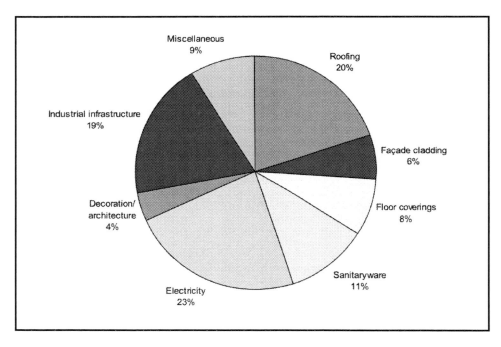

Figure 1.1: *Breakdown of sales of reinforcing fibres in Europe in 1997*

2 Fabrication in fibre-reinforced polymers

A wide range of processing methods is available for FRP composites, although generally the underlying principle is the same. There are differences between the techniques available for thermoset and thermoplastic resins, due to their different properties.

A thermosetting resin, which may be in either a liquid or semi-solid state, is combined with an array of reinforcing fibres and, by the application of heat and/or pressure, the combination is converted to a rigid mass as the resin polymerises. One of the most important features of the manufacture of polymer composites is that the structural material (ie the composite) and the product are formed simultaneously in a single process. Composite structures may be built up *in situ* from the raw materials, as in hand lay-up methods, or they may be made by shaping semi-finished products, in which the components are already combined in the correct proportions.

In the case of reinforced thermoplastics, the fibres are combined with a polymer, which is already polymerized so that the final fabrication is a forming process only. The manner of combining fibres and matrix into a composite material depends on the fibre/resin combination and on the scale and geometry of the structure to be manufactured. The microstructure and properties of the end product will depend on the fabrication process that is chosen in order to meet specific design requirements.

In this section we describe the main features of processes in common use, but it should be noted that there are many variants of these. The methods are divided into seven main categories here.

The terminology or jargon associated with these processes may be a significant barrier to the use of FRP composites. However, the technical terms are described in the **Glossary**, and it is hoped that these descriptions, combined with the diagrams, will provide a reasonable understanding of what is being discussed. Individual processes are not referenced, but further information can be obtained from the text/reference books of Lubin (1982), Bader *et al* (1990), Eckold (1994) and Astrom (1997), all of which are listed in the Bibliography to this Report. Table 5.1 in Chapter 5 presents the different processes in a hierarchical structure and gives the fibre contents that can be obtained from the various processes. This chapter presents the processes in reverse order, starting with the methods that give the highest fibre contents, progressing to lower fibre contents.

2.1   PULTRUSION

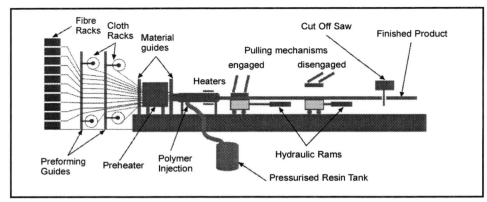

Figure 2.1: *The pultrusion process (SP Systems Ltd)*

In the pultrusion process (Figure 2.1), tightly packed tows of fibres, impregnated with catalysed resin, are pulled through a shaped die to form highly aligned, continuous sections of simple or complex geometry. Curing of the resin may be achieved either by heating the die itself or by the use of dielectric heating.

Solid and hollow sections may be produced by this process, and because of the high fibre content (70 per cent by volume is achievable) and the high degree of fibre alignment resulting from the tensile force used to pull the fibre bundle through the die, extremely good mechanical properties can be obtained (the highest achievable in any variety of composite). Off-axis fibres may also be introduced into the structure if required.

Figure 2.2: *Pultruded profiles (Saint-Gobain Vetrotex)*

Figure 2.3: *Access ladders and walkways from pultruded sections (Strongwell)*

Typical applications of pultruded shapes are concrete reinforcing bars and pre-stressing tendons, I beams and other sections (Figure 2.2), roof trusses, space frames, walkways (Figure 2.3), shear stiffeners, electrification gantries, racking, etc. These racking systems are produced by joining pultruded sections together in the same manner as for timber or steel frameworks.

2.2   FILAMENT WINDING

Cylindrically symmetric structures such as pressure vessels, tanks and a variety of pipes, can be made by winding fibres or tapes soaked with pre-catalysed resin onto expendable or removable mandrels (Figure 2.4).

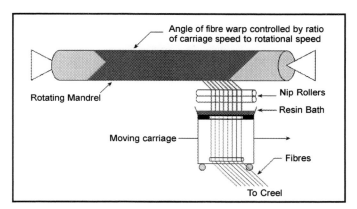

Figure 2.4:   *Filament winding of a cylindrical structure (SP Systems Ltd)*

The process is similar to making cardboard tubes from paper. Non-cylindrical convex shapes can also be made with the same technology, and it can be a competitive technology for making box beams etc. Newer developments allow the on-site reinforcing of columns by over-wrapping with resin-impregnated fibres.

Winding patterns may be simple or complex and can be accurately calculated to resist a prescribed stress system (eg a given ratio of hoop stress to longitudinal stress) in service. Variations in winding pattern, or in the combination of stresses subsequently applied to the structure, will clearly change the extent to which the fibres are loaded purely in tension or to which shear stresses are introduced between separate layers of winding. After the resin has hardened the mandrel is removed and, if size permits, the product may be post-cured at an elevated temperature. Extremely large vessels can be made by this method, but these are usually left to cure at ambient temperature. Since the winding procedure can be closely controlled, a high degree of uniformity is possible in the fibre distribution of filament-wound structures. Cycle times may range from an hour for a small bottle, to many days for a large sewer pipe or pressure vessel.

Figure 2.5: *A filament-wound tank (Saint-Gobain Vetrotex)*

Typical products manufactured by filament-winding are pipes and vessels with a symmetry axis, plates and shaped components, hollow tanks (Figure 2.5) and vessels, including silos. The process of wrapping, eg in retrofit strengthening of bridge piers by overwinding with reinforcing fibres, is an adaptation of the process. Some wind generator blades are produced by a combination of filament winding and resin transfer or infusion.

2.3   COMPRESSION AND TRANSFER MOULDING OF SMC AND DMC

A commercially important class of thermoset moulding compounds, the polyester moulding compounds, consist of chopped glass fibres blended with a pre-catalysed resin and a substantial quantity of inert filler (eg chalk). If the fibres are short (1 cm or 2 cm) a "dough" is produced, which is referred to as a dough-moulding compound (DMC), known in the USA as a bulk moulding compound (BMC).

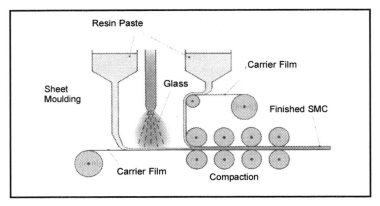

Figure 2.6: *Schematic illustration of an SMC production line (Permali UK Ltd)*

When the fibres are longer (5 cm to 10 cm) a pliable sheet moulding compound (SMC) is produced. Figure 2.6 illustrates the process of manufacturing SMC. DMCs, on the other hand, are produced by blending together the required combinations of resin, chopped fibres, fillers and pigments, and moulding aids in an industrial blender (or mixer) to a thick doughy paste.

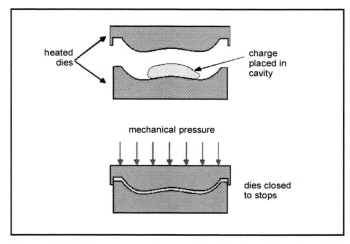

Figure 2.7: *Simple compression moulding of a slug of dough-moulding compound (Bryan Harris)*

Both types of compound can be fabricated to final shape by compression moulding – squeezing to shape between heated, shaped dies (as in Figure 2.7) – or by transfer moulding – forcing a mass of dough from a heated chamber into a shaped die cavity. Careful control of the flow of dough during processing is required to avoid unwanted fibre-orientation effects, which can result in unacceptable anisotropy in some components made from DMC, although this difficulty does not arise with SMCs. Since shaped dies and hot presses are required, production runs of many thousands are necessary if the required shapes are complicated. Suitably adapted injection-moulding presses (see Section 2.5) can also be used to process these materials. The production rates are much higher than for press-moulded DMCs, but the cost is correspondingly greater. Cycle times and the costs of die sets for DMC/SMC mouldings are similar to those referred to above for matched die moulding, although the raw materials are likely to be cheaper.

These methods can be used to produce simple or complex mouldings, domestic services ducting, hot- and cold-water tanks, bath and sink mouldings, rain-water goods, decorative panels and fairings. SMCs and DMCs are used extensively in electrical equipment (Figure 2.8) and in the automotive industry (Figure 2.9).

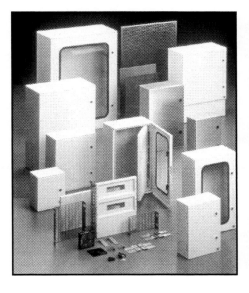

Figure 2.8: *Cabinets, housings etc moulded in SMC (Saint-Gobain Vetrotex)*

Figure 2.9: *The front end of the Renault Espace, manufactured by press moulding of a moulding compound (Saint-Gobain Vetrotex)*

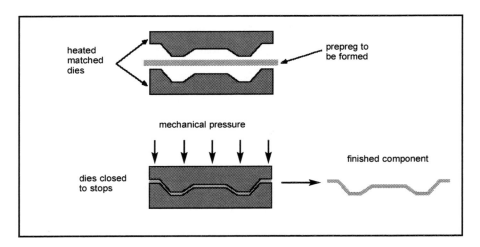

Figure 2.10: *Matched-die moulding (Bryan Harris)*

Large panels and relatively complex open structural shapes are constructed by hot-pressing sheets of pre-impregnated fibres or cloth between flat or shaped platens (Figure 2.10), or by pressure autoclaving to consolidate a stack of prepreg sheets against a heated, shaped die (Figure 2.11). Woven reinforcements are useful for constructing shapes with double curvature since they can be "draped" over complex formers, unlike unidirectional prepreg sheets, which may wrinkle because of their anisotropy. Pressing must be carried out carefully to produce intimate association of the fibres in different layers, with expulsion of trapped air and excess resin. The time/temperature cycle must also be controlled, to ensure final curing of the resin only when these conditions have been met. The orientation of the fibres in the separate layers is varied to suit the required load-bearing characteristics of the laminate or moulding. Composites reinforced with chopped-strand mat (CSM) or continuous-filament mat (CFM) reinforcements may also be press-laminated.

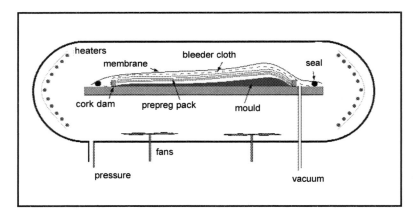

Figure 2.11: *Autoclave moulding (Bryan Harris)*

For the highest quality laminates prepared from prepreg, precision steel dies (male and female in the matched-die process), presses and control equipment are required, and the economics of these processes require production runs of thousands, depending on complexity. A matched die set may cost £100 000. For thermoset mouldings the cycle time is short, it only takes minutes. Press moulding can also be carried out with liquid resins, which can be cured at low temperatures. The required pressures are much lower and the press tools can then be made of composites, or metal-faced composites, and are consequently much cheaper. Cycle times are longer because of the lower conductivity of the composite tooling.

In autoclave moulding, a single shaped die or surface is used and consolidation of the prepreg stack is achieved under heat and pressure in a chamber, which can be both evacuated to remove excess air and pressurised to compact the prepreg and drive out excess resin, as shown in Figure 2.11. Autoclaving dies are less costly than those required for matched-die moulding, since only one shaped die is required, but the pressure vessel and associated control and safety equipment add significantly to the expense.

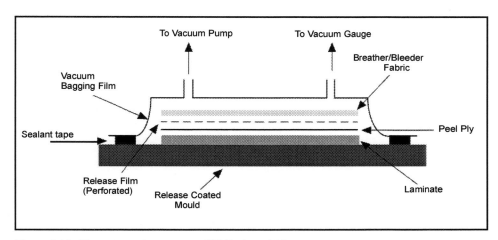

Figure 2.12: *The vacuum-bag process (SP Systems Ltd)*

A much less costly variant of the autoclaving process is to lay up the prepreg stack against a heated shaped surface, enclose the die and stack in a flexible bag and evacuate the bag so as to use atmospheric pressure to consolidate the composite (Figure 2.12). Retrofit strengthening sheets and plates are made by these processes.

2.5   INJECTION-MOULDING OF REINFORCED THERMOPLASTICS

Thermoplastic-based materials, such as glass-filled Nylon, are made by the injection moulding of granules of material, in which the chopped fibres and matrix have been pre-compounded (Figure 2.13).

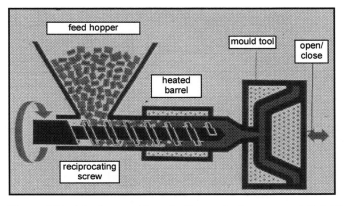

Figure 2.13: *Injection moulding of a thermoplastic compound (Saint-Gobain Vetrotex)*

The flow of material during moulding may be non-uniform, especially in moulds of complex geometry. Preferential fibre orientation, which can be achieved by appropriate gating design, is often useful but only if it can be properly controlled. This process requires very expensive dies and presses and is traditionally associated with the repetitive manufacture of large numbers of relatively small components, with production runs of approximately 100 000. Cycle times vary between a minute and a few minutes, depending on size and complexity. Die costs are greater than those for the processing of thermoset composites, because of the necessity of building in oil-cooling or water-cooling channels to maintain the die temperature below the melting temperature of the matrix polymer. A die set of the level of complexity of a milk-bottle crate, for example, could cost £300 000 or more. The dies and presses used need to be robust because of the high closing pressures in this kind of moulding.

Figure 2.14: *A washing machine drum injection-moulded in reinforced thermoplastic (Saint-Gobain Vetrotex)*

The method is used for door furniture, water fittings, hot-water and cold-water tanks, electrical fittings, pump bodies and impellers, and equipment housings. Substantial use of these materials is also made in the automotive industry, even in relatively hot applications and domestic appliances (Figure 2.14). Sheets of reinforced thermoplastics are also post-formed (eg to make baths, sinks etc) by hot pressing or stamping.

2.6   CONTINUOUS SHEET PRODUCTION

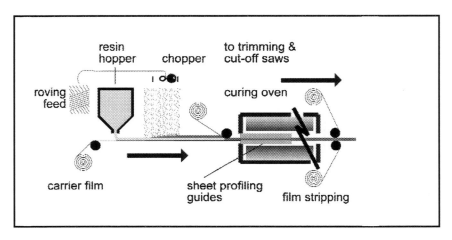

Figure 2.15: *Continuous sheet production (Saint-Gobain Vetrotex)*

For continuous sheet production chopped strand mat or chopped strands are impregnated with resin and sandwiched between two layers of film on a moving belt. The sandwich passes through guides that form the corrugated, or other desired, profile, as it enters a long oven section in which cure takes place. On leaving the oven, the processing films are stripped off, the sheets trimmed and finally the completed panels are cut to length by a flying saw.

2.7   RESIN-TRANSFER MOULDING (RTM) AND VACUUM-ASSISTED RESIN-TRANSFER MOULDING (VARTM)

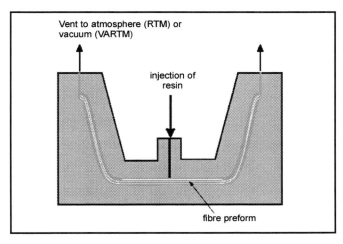

Figure 2.16: *Resin-transfer moulding (RTM), showing optional vacuum connection for VARTM (Bryan Harris)*

RTM is a process for producing high-quality mouldings of complex shape at much lower cost than with matched die moulding. Pre-catalysed resin is pumped under low pressure (1 to 4 bar) into a fibre preform, which is contained in a closed (and often heated) die (Figure 2.16). The preform may be made of any kind of reinforcement, but usually consists of woven cloths or continuous-filament mats, arranged as required by the design. Thick components containing foamed polymer cores can also be produced in this way. In high-speed versions of the process, favoured by the automobile industry, the cycle time may be reduced to a few minutes, but in normal operation a cycle time of 15 to 30 minutes, depending on the resin cure characteristics, is easily achievable. The mould cavity is often evacuated to assist wetting out, when the process is referred to as vacuum-assisted RTM or VARTM.

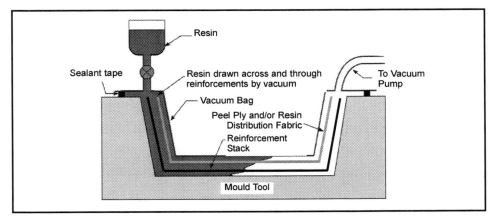

Figure 2.17: *The resin-infusion process, or soft-tool RTM (SP Systems Ltd)*

Since RTM is a low-pressure process, the equipment is relatively inexpensive and the dies can be made from easily shaped materials, including metal sheet, electro-formed nickel, cast or machined aluminium alloys, or GRP, although these must be backed with rigid structures to prevent distortion during mould filling. As a consequence, this is an important, high-volume, automatable process, which offers considerable advantages in the production of low-cost engineering components, when used in association with automated fibre-preform manufacture. In the construction context, however, it is a medium-rate production process (100 to 2000) which also has the capability of producing small numbers of complex shapes (10 to 100) economically. A wide range of low-viscosity resins – polyesters, vinyl esters, epoxies, urethane-acrylates, etc – are now commercially available for use in this process.

The process of resin-infusion is effectively an extension of RTM, in which a pre-arranged fibre pack is laid against a single-sided shaped die and covered with a flexible sheet. The space is evacuated and atmospheric pressure consolidates the fibre pack, whereupon pre-catalyzed resin is then bled into the fibres to form the desired component (Figure 2.17). On account of the low tooling costs this is a cheap process and can be used effectively for the economic manufacture of low production volumes (one or two components to one or two hundred).

Figure 2.18: *Translucent cladding panels manufactured by resin infusion (Saint-Gobain Vetrotex)*

Figure 2.19: *Large-capacity water storage tank. Tank panels are easy to handle even on difficult terrain with difficult access (Permali, UK, Ltd)*

Cladding and translucent roofing panels (Figure 2.18), large tank panels (Figure 2.19), sculptures, folded-plate and shell structures, complete bridge decks, and a wide variety of other structural components with varying degrees of anisotropy/orthotropy can be produced by these resin-transfer methods.

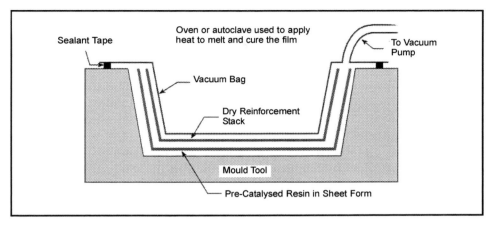

Figure 2.20: *The resin film infusion process (SP Systems Ltd)*

In another variant of the resin-transfer process, a solid film of the pre-catalysed resin is enclosed in the mould, together with the reinforcement preform (Figure 2.20). The evacuated stack is then placed in an oven so that the film melts, infuses the preform and cures, in a single operation.

2.8 OPEN-MOULD PROCESSES: CONTACT MOULDING BY HAND LAY-UP OR SPRAY-UP

Composites reinforced with chopped-strand mat (CSM) or woven cloth are often laid up by hand methods, especially for irregularly shaped structures. Large structures like tanks, boats and pipes are made in this way. A shaped former or mandrel is coated with a gel-coat and the required shape and thickness are then built up by rolling on layer upon layer of resin-impregnated cloth or mat (Figure 2.21). A final flow coat is then applied and the finished structure may be post-cured (by heating).

Alternatively, after application of the gel-coat to the mould, the structure may also be built up by spraying through a gun which simultaneously delivers short fibre bundles (chopped in the gun) and pre-catalysed resin (Figure 2.22). The distribution of fibres in such structures will depend on the skill of the operator, and if there is poor quality control, resin pockets and voids may be present in these materials.

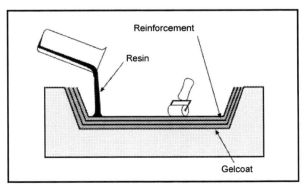

Figure 2.21: *Schematic illustration of wet or hand lay-up (SP Systems Ltd)*

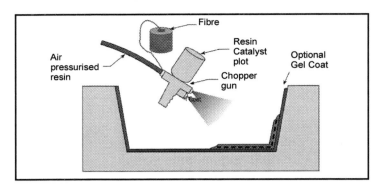

Figure 2.22: *The co-spraying of resin and chopped fibres (SP Systems Ltd)*

The process is suitable for small numbers of components (1 to 100). The moulds for this process can be made cheaply (£10 to £1000, depending on size and complexity) from materials such as timber, plastic foam, metal sheet, or from GRP. It is also possible to fabricate one-off structures, in which the former actually becomes part of the finished article. Cycle times may range from a few hours to days, again depending on size and complexity.

Figure 2.23: *Manufacture of a bath by contact moulding (hand lay-up) (Saint-Gobain Vetrotex)*

Typical components manufactured by these processes include decorative shapes (eg made-to-order sculptures, finials, etc), folded-plate structures, shell structures, beams, concrete formwork, boats and similar open vessels or containers, and shapes for architectural repairs. Complete, ready-to-install units, like bathrooms with washbasins and bath moulded into a one-piece structure, have also been produced (Figure 2.23).

2.9 DEFECTS IN MANUFACTURED COMPOSITES

All practical reinforced polymers are likely to contain defects of various kinds, arising from the processes of manufacture, and their mechanical properties will therefore exhibit some degree of variability. This variability will be greater in materials produced by hand lay-up methods than in composites made by mechanised processes. The nature and severity of defects found in a manufactured product will be characteristic of the manufacturing process

(Harris and Phillips 1983, 1990a). The defects that may be present in manufactured composites include:

- incorrect state of resin cure, eg resulting from variations in local exotherm temperatures in thick or complex sections during autoclaving or hot pressing
- incorrect or variable fibre volume fraction, often with local resin pockets
- pores or voids in resin-rich regions
- misaligned or broken fibres
- gaps, overlaps or other faults in the arrangement of plies
- disbonded interlaminar regions (delaminations)
- resin cracks or transverse-ply cracks resulting from thermal mismatch stresses
- local bond failures in adhesively bonded composite components

Any of these defects can locally reduce the strength of the material below the required component design stress and they may also act as sites for the initiation of damage during subsequent monotonic or cyclic loading. Non-destructive evaluation (NDE) methods are available for the assessment of the distribution and severity of defects, as described in Chapter 5.

3   Case studies

This section shows a range of current uses of polymer composites in construction and a number of representative examples from other industries. These are included to enable technology transfer to occur from industries that are making greater use of FRP composites.

The examples are intended to give a balance to each of the following:

- application type
- material used
- manufacturing process used
- reason for the use of FRP composite material.

Although there are inevitably gaps, each case study has tried to provide the following information:

- name, place and date
- the nature of the application
- why it was made with FRP composite
- current condition
- maintenance history
- source(s) of further information
- cost information (note that these are generally from the time of construction, and vary in terms of what the price refers to – the best available information is used).

The reasons for the use of FRP composites have been divided into nine main groups:

1 *Time saving* – low weight for fast construction and minimum disruption.
2 *Durability* – able to survive, especially in harsh environments.
3 *Repair* – to allow the repair of structures *in situ*.
4 *Strengthening* – to allow the strengthening of structures *in situ*.
5 *Lightweight* – where especially high performance is needed in one direction.
6 *Appearance* – where a particular colour, shape or texture is required.
7 *Blast/fire* – resistance to blast or fire.
8 *Radio* – transparency to electromagnetic radiation.
9 *Low maintenance* – in conditions where difficult access makes maintenance difficult.

The examples are grouped by application type, to make it easier for the reader to find particular examples relevant to their concerns.

Table 3.1: *List of case studies and key design reasons for use of FRP composites*

Name	Country	Location	Time saving	Durability	Repair	Strengthening	Lightweight	Appearance	Blast/Fire	Radio transparent	Maintenance	Case study no
Buildings												
Wing tower	UK	Glasgow					✓	✓				3.1.1
Eyecatcher building	CH	Basel						✓				3.1.2
Roof Trusses	UK	Darvel, Ayrshire					✓					3.1.3
Stone Mountain building	USA	Georgia							✓			3.1.4
Burj Al Arab Hotel	UAE	Dubai						✓				3.1.5
Radomes	various									✓		3.1.6
Second Severn Crossing Visitors Centre	UK	Avon	✓	✓								3.1.7
Mobile phone headquarters	South Africa							✓		✓		3.1.8
Large structural frame	Middle East									✓		3.1.9
Wiltshire radio station	UK	Wiltshire	✓							✓		3.1.10
American Express building	UK	Brighton		✓				✓				3.1.11
Herman Miller factory	UK	Bath		✓				✓				3.1.12
New Covent Garden roof	UK	London		✓			✓	✓				3.1.13
Bridges												
Kolding Bridge	Denmark						✓					3.2.1
Fidgett's footbridge	UK	Oxfordshire					✓					3.2.2
Parsons Bridge	UK	Wales		✓			✓					3.2.3
Bonds Mill Lift bridge	UK	Stroud, Gloucs					✓					3.2.4
Shank Castle footbridge	UK	Cumbria		✓			✓					3.2.5
Aberfeldy footbridge	UK	Aberfeldy	✓				✓				✓	3.2.6
Repair / Strengthening												
Cleddau bridge	UK	Wales				✓						3.3.1
Office repair	UK					✓						3.3.2
Tickford bridge	UK	Newport Pagnell				✓						3.3.3
Sins bridge	Switzerland					✓						3.3.4
Ibach bridge	Switzerland				✓							3.3.5
Liverpool Cathedral	UK	Liverpool						✓				3.3.6
Water												
Gratings for drainage products	UK			✓								3.4.1
Sewage tank covers	UK	Horsham		✓			✓					3.4.2
Water tanks	UK			✓								3.4.3
Other												
Wind wand	UK	London						✓				3.5.1
Cliff stabilisation	UK	Devon					✓					3.5.2
Rest zone	UK	Greenwich	✓									3.5.3
MOMI hospitality tent	UK	London					✓					3.5.4
Carwash	UK	Hornchurch		✓								3.5.5
M25 Message gantries	UK	London		✓							✓	3.5.6
Blades for wind turbine	various			✓			✓				✓	3.5.7
Lifeboat hull	UK			✓			✓				✓	3.5.8

3.1   BUILDINGS

FRP composite materials have been widely used in building applications for more than 30 years, with the most common application being cladding, usually GRP panels. Many of these examples have survived and are in excellent condition – some of these are given in the case studies that follow. There were also problems with some of the panel systems produced, and this lowered confidence in the material as a whole, leading to a reduction in the use of GRP cladding. However there is no reason why this cannot be a successful application for FRP composite materials, giving a versatile visual appearance, with a long life and limited need for maintenance.

3.1.1 Glasgow Wing Tower, 2001

What? A landmark tower for the city of Glasgow situated at the end of the Pacific Quay on the banks of river Clyde.

A viewing cabin is on the top of the only known rotating tower in the world. The cabin is 104 metres above ground and accommodates 24 people at a time, reached via external lifts. The tower will be driven to face the wind by means of four electric motors, which are fixed below four rollers.

Figure 3.1: *The cabin being installed on the wing tower (Buro Happold)*

At the top of the tower is a 24.5 m carbon fibre mast, making the total height 134 m.

How / what materials? The main steel structure consists of a shaft, which contains the stairs and airfoil-shaped outriggers to meet the requirement of aerodynamic design. The viewing cabin is made in GRP, and has a volume of 78 m³. It consists of a composite panel having overall thickness of 23 mm and comprising a 20 mm thick core of foamed PVC. This has a density of 65kg/m³ with a glass tissue faced, fabric reinforced, polyester resin laminate bonded to both faces, giving a 1-hour fire resistance.

The mast is essentialy rectangular in section, reducing in size from 1000 mm x 500 mm down to 300 mm x 150 mm. It is made of a carbon fibre composite, with Young's Modulus E = 7000 N/mm².

Figure 3.2: *The mast (Buro Happold)*

Why? The cost of GRP was less and the structure lighter in weight when compared to an aluminium alternative. The tender return showed that by using GRP the cost was reduced by 60 per cent.

For the mast, only carbon fibre could produce the required stiffness and strength for this height, steel could only have achieved 10 m.

Current condition: Just built

Cost: £350 000 for the cabin, £105 000 for the mast.

3.1.2 The Eyecatcher building, Basel, Switzerland, 1999

What? The Eyecatcher Building is a 15 m high, 600m² fully-composite office building. It was constructed for the Swissbau 99 Fair and later disassembled and relocated in the centre of Basel.

How / what materials? Standard GRP I- and U-profiles.

Why? FRP composites were chosen as a demonstration, to show possibilities of using non-conventional construction materials. The lightness of the building allowed easy disassembly and relocation at the end of the Fair.

Current condition: Good

Maintenance history: No maintenance yet

Further information from: IABSE website:
www.iabse.ethz.ch/sei/backissues/abstracts.sei9904/keller

Cost: Not known

Figure 3.3: *The Eyecatcher house, close up of structure (Fibreline Composites A/S)*

Figure 3.4: *The Eyecatcher house (Fibreline Composites A/S)*

3.1.3 Roof trusses over reservoir, Darvel, Ayrshire, Scotland, 1998

What? 19 m long trusses to support the roof over a water storage reservoir in the Irvine valley.

How / what materials? All-composite structure fabricated from standard structural pultruded GRP sections. Bolted connexions are required to resist axial loading (top figure).

Why? This is a lightweight structure, easily erected in the field by conventional construction methods, but without the need for the heavy handling equipment that would be needed for a steel construction.

Current condition: No information

Maintenance history: None

Further information: Rosner C N, Rizkalla S H, 1992, Design of bolted connections for orthotropic fibre-reinforced composite structural members, in *Advanced Composite Materials in Bridges and Structures* (Neale K W and Labossière, Editors) (Canadian Society for Civil Engineering, Montreal, Canada) 373–382

Anon, 1995, *Fibreline Design Manual for Structural Profiles in Composites* (Fibreline Composites A/S, Kolding, Denmark)

Gilby J, 1998, Pultrusion provides roof solution, *Reinforced Plastics* 42 (6), 48–52

Turvey G J, 2000, *Bolted connections in PFRP structures*, 2, 146–156

Cost: Not known

Figure 3.5: *Roof trusses over the Darvel reservoir (Composite Solutions Ltd)*

Figure 3.6: *Detail of bolted connection in roof trusses of Figure 3.5 (Composite Solutions Ltd)*

3.1.4 Stone Mountain, Georgia, USA, 1996

What? The building is the Aerial Tram Station for Stone Mountain, part of the cable car system for the Stone Mountain Park. It was extensively refurbished, including the replacing of the roof, and then used for the 1996 Winter Olympics.

How / what materials? The roof structure consists of pultruded I beams in GRP up to 12.2 metres long and 60 cm deep. These are joined together using GRP studs and nuts. The structure is clad with a pultruded GRP cladding system.

There are 64 of the large I beams working as rafters, and 17 columns, with a total of 130 hand lay-up connection plates. The cladding consists of GRP pultruded over a foam core to give lightweight but strong insulation panels.

Why? There are three reasons:

1   The roof space contains large quantities of broadcasting equipment
2   It gives the structural strength required
3   It meets the visual demands of the project team.

Current condition: Not known

Further information from: Not known

Cost: Not known

Figure 3.7:
Cladding being installed (Strongwell)

Figure 3.8:
Roof structure (Strongwell)

3.1.5 Burj Al Arab, Dubai, United Arab Emirates, 1994

Figure 3.9: *Burj Al Arab (Peter Jennings ACO Technologies)*

What? Burj Al Arab is a luxury hotel, soaring 321 m above the Arabian Gulf, and stands on a man-made island, 280 m offshore. The billowing sail shaped structure is designed to resemble a modern sailing boat. Construction involved the use of 2400 m² of cladding. The sail façade is constructed from a double-skinned Teflon-coated woven glass fibre used as a tensioned membrane.

How / what materials? Both the cladding panels of the atrium wall and the balcony upstand are constructed from GRP. The cladding panels are constructed with a double skin separated by stiffening ribs and rigid board insulation. The three-dimensional composite panels cover 35 000 square metres. They were manufactured in a specially set up small factory unit within close proximity of the site.

Figure 3.10: *Interior of the hotel (Jumerah International)*

Why? The dominant reason was to achieve the visual design sought by the architect.

Current condition: Excellent (both internal and external)

Further information from: *Reinforced Plastics,* June 2000 (Elsevier Science Ltd) p 7

Cost: Not known

Figure 3.11: *The hotel at night (Jumerah International)*

3.1.6 Radomes, world-wide

What? Self-supporting, light-weight structures to contain communications equipment, very often in remote locations, including high altitudes and high latitudes. Designed for wind speeds up to 120 m/s (270 mph) at -50°C, and to withstand extreme conditions of rain, ice and heavy snow falls.

How / what materials? These are monocoque shells which may be from 5 m to 30 m in diameter, and are frequently made of glass- and/or aramid-fibre-reinforced composite sandwich panels. They are produced by resin injection in closed moulds so that both inner and outer surfaces have a high-quality finish, or may be produced by conventional contact moulding.

Figure 3.12:
An 11.6 m dia radome located in Serbia (W & J Todd Ltd)

Why? These shells are designed to be free of any internal framework that might interfere with electromagnetic radiation. They are designed, for both civil and military purposes, for optimum electrical performance within the L and S band dual-frequency range. They are assembled from separate panels, which can be easily assembled on site, even in remote locations where handling equipment is not available. The intrinsic strength of the geometrical form and the method of manufacture makes for highly efficient and cost-effective structures. The foam core in the sandwich panels provides thermal insulation in addition to improving the rigidity and stability of the structure.

Current condition: Radomes have been in use for many decades and show no deterioration in condition or performance.

Figure 3.13:
Radome for air defence, located across Europe 1995 (CETEC Consultancy Ltd)

Further information from: Kendall D, 1996, *Composites in Communications*, Proceedings of 20th BPF Composites Congress Sept 1996, The British Plastics Federation, London

Kendall D, 1999, *The structural use of moulded composites: applications and opportunities*, Proceedings of a conference on Composites and Plastics in Construction, Nov 1999, BRE, Watford, UK, (RAPRA Technology, Shawbury, Shrewsbury, UK), paper 19

Cost: Approximately £250 000 inclusive for a 25 m diameter dome, including assembly.

3.1.7 Second Severn Crossing visitor centre, UK, 1992

What? At the Second Severn Crossing between England and Wales, the double-storey site offices were constructed from all-composite materials in 1992. There is space for eight housed staff, and a 20-person meeting room.

How / what materials? Pultruded interlocking GRP panels were bonded together using epoxy adhesive in order to produce a membrane structure in which walls, floors and roof all provide the necessary structural integrity.

Figure 3.14:
Second Seven Crossing visitor centre (BRE)

Why? GRP was chosen on account of its rapid erection time (which in turn is because it is lightweight). Cost savings were made because there was no need for an additional framework. Good thermal insulation and fire-resistance are key features of the building design. Panels are insulated to a high standard and double-glazed windows have been fitted. In critical areas, panels meet Class 0 fire regulation specifications with a fire resistance of at least 30 minutes when tested in accordance with BS 476: Part 21. The Advanced Composite Construction System provides rapid, cost-effective, high-quality construction, even in locations where access is difficult. The key advantages from using composites in this application are:

- monocoque structure, no additional framework required
- lightweight modular components
- designed for easy erection
- thermally insulated and fire resistant.

Figure 3.15:
The visitor centre (Maunsell Ltd)

Current condition: BRE inspected the building in 1999. The structure was still in very good condition.

Maintenance history: The building changed use in 1998, including considerable internal changes, to become the visitor centre. Not aware of any external maintenance to date.

Cost: £75 000

3.1.8 Moblie phone company headquarters, South Africa, 1998

What? Cladding panels for the headquarters of a South African mobile phone company. The appearance matches metallic cladding systems.

How / what materials? The panels are hand laid up GRP.

Why? Composite panels provided the advantage of electronic transparency and were a cost-effective alternative to other forms of cladding with similar appearance.

Figure 3.16:
Composite cladding panels on moblie phone company headquarters (Ove Arup)

Current condition: Good

Maintenance history: No information

Further information: None available. For further information on electro-magnetically transparent structures, see 3.1.6 Radomes, 3.1.4 Stone Mountain and 3.1.9 Large scale frame.

Figure 3.17:
Composite cladding panels on moblie phone company headquarters (Ove Arup)

Cost: Not known

3.1.9 Large structural frame, Middle East

What? A large structural framework, 9 m x 11 m x 10 m high, to support sensitive electrical equipment.

How/what materials? The frame is produced from a variety of pultruded GRP sections using vinylester resins, the largest being 30 cm x 30 cm x 1.25 cm flange beams. The connections were made using threaded GRP studs with moulded nuts, and bonded with a structural 2-part epoxy adhesive once erected. Very tight tolerances were required, therefore an allowance was made in the joints for adjustment based on measured deviations of the manufactured sections.

Figure 3.18: *GRP pultruded frame (CETEC Consultancy Ltd)*

Why? The primary design criterion was that the frame should be non-magnetic and non-conducting throughout, including the fasteners.

Current condition: Good

Maintenance history: No information

Further information: None available. For further information on electro-magnetically transparent structures, see 3.1.6 Radomes, 3.1.4 Stone Mountain and 3.1.8 Mobile phone company headquarters, see also Figure 4.26.

Cost: Not known

Figure 3.19: *Wiltshire radio station, general view (Don Gray)*

3.1.10 Wiltshire Radio Station, Wooton Bassett, 1982

What? A radio station building in the gardens of a Grade II listed manor house. It comprises six studios and a news reporters' room.

How / what materials? The structure comprises concrete blockwork clad with acoustic GRP panels, which are filled with sand. The panels are stiffened by a corrugated pattern on their outer faces; this form also prevents the sand from causing bulging. The joints between the panels are sealed with neoprene extrusions.

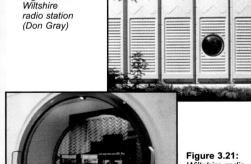

Figure 3.20: *Wiltshire radio station (Don Gray)*

Why? The building had stringent requirements in terms of acoustic performance. This was achieved by filling the GRP panels with sand. The architectural form of the building was also important as was a design that could be achieved within the budget and on-air deadlines.

Current condition: Not known

Maintenance history: Not known

Further information from: Amery, Colin (1995) Architecture, Industry and Innovation: *The early work of Nicholas Grimshaw & Partners*, Phaidon *Building Magazine*, 5th August 1983

Figure 3.21: *Wiltshire radio station close up (Don Gray)*

Cost: Contract value £198 000 for 170 square metres.

3.1.11 American Express Building, Brighton, 1977

What? Amex headquarters, intended to be a brilliant white building, in keeping with established Regency tradition, to provide maximum office accommodation without prejudicing daylight to adjacent properties.

How / what materials? Single-skin contact-moulded (hand lay-up) CSM GRP beam-panels up to 10 m long containing sandwich and tubular areas and top-hat internal members (pre-formed). Gel coat of optically stabilized, fire-retardant white isophthalic polyester resin. Material fully post-cured to ensure structural and dimensional stability.

Figure 3.22: *The American Express building, Brighton (Scott Bader Co Ltd)*

Why? The building was required to resist the marine environment over a minimum service life of 60 years. GRP was considered to be the only material that would provide a permanently white finish despite conflict between colour-fastness and fire rating.

Current condition: After 24 years, GRP is in excellent condition and has weathered to a semi-gloss finish with little or no yellowing, Some evidence of star cracking in the gel coat at corners and highly stressed regions (< 0.1 per cent of surface). No evidence of blistering or lifting of the gel coat. Building lost roof in 1986 hurricane, but no adverse effect on panels or fixings.

Maintenance history: Mild chemical cleaning treatment has restored some gloss.

Further information from: T Starr, *Reinforced Plastics*, June 1998, (Elsevier Science Ltd) 42–46

Cost: £15 million

3.1.12 Herman Miller Factory, Bath, UK, 1976

What? The structure of the building is a simple primary and secondary beam system with columns on a 10 m x 20 m grid. The cladding system is completely demountable, enabling the panels and glazing to be interchanged by unskilled labour, to allow the staff to alter the building to suit changing needs.

Figure 3.23: *The Herman Miller Building, Bath (Bryan Harris)*

How / what materials? Panels were produced with two separate sandwich skins, the function of the core being to keep the two skins at a constant distance apart during deflection and keep them rigidly connected so they acted as a composite element. A neoprene gasket joint is used for both the vertical and horizontal joints.

Why? The primary reason is the visual appearance that the project team desired, combined with the anticipated long life time and limited maintenance requirement.

Figure 3.24: *The good condition of the surfaces of the cladding panels can be seen by the degree of reflectivity (Bryan Harris)*

Current condition: Excellent smooth surface condition. From ground level, no visible signs of chalking, cracking or blistering, and colour changes are imperceptible. No rainwater run-off discoloration.

Further information from: Brookes A J (1998) *Cladding of Buildings* (E&FN Spon, London)

Herman Miller Factory at Bath, *Architects' Journal*, 1 Mar 1978

Amery C, (1995) *Architecture, Industry and Innovation – The Early Work of Nicholas Grimshaw & Partners* (Phaidon Press Ltd, London)

Action Factory, *RIBA Journal*, Sept 1977

Goldstein B (1978) *Trim-Tech, Progressive Architecture*

Cost: Total contract value: £ 850 000

3.1.13 New Covent Garden Flower Market Roof, 1972

What? GRP cladding forms the complex geometry of the roof cladding. The roof was required to perform well in aspects such as thermal insulation, natural lighting, solar reflectivity and weather tightness.

How / what materials? 100 m² roof cladding in GRP. The design comprises repetitive forms of truncated pyramids on a 4 m grid. The base of the pyramid is translucent. The units were formed through an injection process, which produced a double-skinned unit in one operation. After overcoming a problem of air inclusions in the early stages of production a high standard of surface finish and structural integrity was achieved. To meet fire requirements a smoke baffle arrangement was created within the roof and an external drencher system was installed. Each cell of the roof grid has individual drainage.

Why? Combination of visual appearance, light transmission and low maintenance requirements.

Current condition: Not known

Maintenance history: An inspection in 1982 (reported by Leggart) showed a number of cracks thought to be caused by slight variations in manufactured thickness with consequent resin-rich areas. Otherwise the roof was considered to be in good order. Leggart also reports that problems had arisen in places due to failed jointing, however several feared problems had not occurred such as an obscuring of the panels by dirt and collected debris.

Further information from: Leggart, A J (1984) *GRP and Buildings* (Butterworth, London)

Figure 3.25
New Covent Garden flower market roof (Scott Bader Co Ltd)

Figure 3.26:
Installation of a module (Scott Bader Co Ltd)

Figure 3.27:
Roof showing joints at ridges (Scott Bader Co Ltd)

3.2 BRIDGES

In recent years there have been a number of bridges built with FRP composites as the primary, if not the only, material being used. This is because of the combination of resistance to the external environment, and low weight giving ease of installation and use (in the case of moving bridges). The cost and ease of installation can outweigh the generally higher material costs of the FRP composite solution, particularly in some of these examples for smaller bridges in less accessible locations.

In time there may be more substantial bridges made by using the same techniques, as confidence in their effectiveness grows. There are many potential applications in the rail sector, where charges for line occupation time is often the dominant cost.

3.2.1 The Kolding Bridge, Denmark, 1997

What? The Kolding Bridge is Scandinavia's first composite bridge. It is a pedestrian and cycle bridge, crossing a railway line.

How / what materials? Standard GRP panels.

Why? The busy railway line restricted installation work to only a few hours during nights between Saturdays and Sundays. Installation was completed in just three nights, substantially quicker than would have been possible using more conventional construction materials due to lightness of the GRP. The total consumption of energy required for raw materials, production and assembly accounted for only one quarter of the would-be energy consumption for a similar construction in steel or concrete. Maintenance expected to be cosmetic only for 50 years.

Current condition: Good

Maintenance history: No maintenance yet

Further information from: The Fiberline bridge: A break-through in composite design technology [Anon] *Materials & Design* 18: (1) 43 1997

Cost: Not known

Figure 3.28:
The Kolding Bridge being delivered (Fibreline Composites A/S)

Figure 3.29:
The Kolding Bridge (Fibreline Composites A/S)

Figure 3.30:
The Kolding Bridge (Fibreline Composites A/S)

Figure 3.31:
The Kolding Bridge being installed at night (Fibreline Composites A/S)

3.2.2 Fidgett's footbridge, Oxfordshire, UK, 1995

What? The first concrete footbridge to be fully reinforced with FRP composite rods in the UK. It consists of a slab 5 m long by 1.5 m wide with a depth of 300 mm.

How / what materials? Reinforcement consists of 13.5 mm diameter rods at 150 mm centres in both directions at the top and bottom surfaces. The slab was precast with a 40N/mm² concrete and installed on mass concrete abutments. Vibrating wire strain gauges and thermistors were cast into the concrete and fibre-optic sensors fitted to the slab. The bridge was fitted with GFRP handrails.

Why? The concrete bridge was reinforced with GRP bars as a demonstration project as part of the EUROCRETE programme. It is seeking to test the effectiveness and durability of this solution.

Maintenance history: The structure was completed in 1995 and load tested to 1.25 times design load in accordance with BS8110 and then monitored for the following year.

Further information from: Clarke J and O'Regan P The UK's first footbridge reinforced with glass-fibre rods, *Concrete*, July/August 1995

Parker D, Plastic Surgery, *New Civil Engineer*, 18 May 1985

Weaver A, Bridge tests durability of rebar, *Reinforced Plastics*, July/August 1995 (Elsevier Science)

Figure 3.32:
Fidgetts footbridge (John Clarke)

Figure 3.33:
Reinforcing bars as concrete is poured (Fibreforce Composites Ltd and Eurocrete Ltd)

3.2.3 Parson's Bridge, Dyfed, Wales, 1995

What? Parson's Bridge is a footbridge, which crosses a gorge in the Afon Rheidol in Wales. It is used by ramblers, tourists and locals (who attend a nearby church). It is all-composite, single-span, 17.5 m long, 0.76 m wide and it weighs 1 tonne.

How / what materials? It is constructed of GRP panels, bonded together with epoxy adhesive according to the Advanced Composite Construction System (ACCS). The entire bridge was airlifted into position.

Why? Traditional materials would have made the bridge considerably heavier. This would have necessitated the use of a heavy freight helicopter to carry out the lift. The use of a lighter all-composite structure meant that a much cheaper local helicopter could be used instead.

Current condition: Good structural condition, but (as for other materials) due to the shaded, damp nature of the site, there is some moss growth on the bridge.

Maintenance history: No significant work to date

Cost: £30 000

Figure 3.34:
Parson's Bridge (Maunsell Ltd)

Figure 3.35:
Parson's Bridge (Maunsell Ltd)

3.2.4 Bonds Mill Lift Bridge, Gloucestershire, UK, 1994

What? Bonds Mill Lift Bridge is the first all-composite road and lift bridge in the world to carry traffic loading. It bridges the Stroudwater Navigation Canal by means of a lifting structure. It is 8.5 m long by 4.4 m wide and has a mass of approximately 4.5 tonnes.

How / what materials? Pultruded GRP panels were bonded together with epoxy adhesive. UV inhibitors were added to the resin to increase resistance to outdoor exposure.

Why? GRP was chosen principally because it is lightweight. This meant that no lifting tower or counterweight was necessary and the foundations from the previous fixed-bridge could be used with little modification. This led to significant savings. The corrosion resistance of GRP also meant that durability could be ensured in this damp environment.

Figure 3.36:
Bonds Mill Lift Bridge (Maunsell Ltd)

Current condition: In 2000 the condition of the bridge was found to be good, especially the top wearing surface.

Maintenance history: In 1998 roadway panels with manufacturing defects were replaced.

Further information from: Busel J P, Lindsay K (1997) On the road with John Busel: A look at the world's bridges, *CDA/Composites Design & Application*, Jan/Feb, 14–23 (also on:iti.acns.nwu.edu/composites/cda.html)

Cost £90 000

3.2.5 Shank's Castle Footbridge, Cumbria, UK, 1993

What? Shank's Castle Footbridge crosses the Rae Burn River in a wooded area, about 10 miles north of Carlisle. The bridge is all-composite, single-span, 12 m long, 0.76 m wide and weighs just 500 kg.

How / what materials? Pultruded GRP panels are bonded together with an epoxy adhesive. The bridge was transported by road as far as possible, and then a trolley was used for transportation through the wood. Ropes and winches were used to lift the bridge onto its abutments.

Why? GRP was chosen over timber because of its durability, low weight and ease of construction, leading to time savings.

Current condition: Good

Maintenance history: No significant work to date

Cost: £6000

Figure 3.37:
Shank's Castle footbridge (Maunsell Ltd)

3.2.6 Aberfeldy Footbridge, Scotland, UK, 1992

What? Aberfeldy Footbridge was the first major advanced composite footbridge to be built in the world. It is a cable-stayed bridge and spans 120 m overall across the River Tay on the Aberfeldy Golf Course.

How / what materials? Pultruded interlocking GRP panels were bonded together with epoxy adhesive in order to produce the deck. The deck is then stayed from 18 m-high GRP A-shaped pylons with sheathed Parafil ropes (Kevlar). Construction was carried out by Dundee University final-year students over a 10-week period without the use of cranes.

Why? GRP was chosen because it is lightweight. This meant that construction was carried out quickly and without expensive craneage. The limited need for maintenance was also important.

Maintenance history: Limited

Further information from: Burgoyne C J, 1993, Aberfeldy Bridge: an advanced textile reinforced footbridge, *Proc TechTextil Symposium, Frankfurt*, paper 418

Harvey W J, 1993, A reinforced plastic footbridge, Aberfeldy UK, *Structural Engineering International* 4 229–232

Cost: £250 000 for whole bridge (estimated).

Figure 3.38:
Aberfeldy Bridge (Maunsell Ltd)

Figure 3.39:
Aberfeldy Bridge (Maunsell Ltd)

3.3 REPAIR AND STRENGTHENING OF BRIDGES AND BUILDINGS

There are many hundreds of examples worldwide of concrete structures that have been strengthened, repaired or retrofitted using FRP composite materials. The largest number of these are bridges, although buildings and chimneys have also been strengthened. This application of the use of FRP composites was first used in the bonding of GRP plates on the Kattenbusch bridge in Germany in 1986. In 1991, the use of CFRP sheets was introduced in Switzerland, when the concrete Ibach Bridge was strengthened.

It is not only concrete structures that have been strengthened with FRP composite materials. Timber and iron structures have also had their carrying capacity substantially enhanced by the addition of CFRP sheet.

The average annual expense for the repair of bridges alone in Europe exceeds £2 bn (Meier 1995). Therefore, the need for a cheap, reliable and simple retrofit solution is clear. Since about 1980, bonded steel plates have been used as strengthening components on bridges and buildings. However, they have possible corrosion problems and are heavy, which makes bonding them to existing structures rather awkward and therefore expensive for all but the shortest lengths of plate. In fact, if bonding of plates is required on the inside of a box-section concrete bridge, for example, the plates would have to be man-handled within the box, which would preclude the use of long, wide or thick steel plates. This also brings significant health and safety challenges with it, which are eased for CFRP as it is delivered in rolls that are much more readily applied.

Another successful application of strengthening and repair is that of wrapping concrete bridge piers with FRP composite sheet or tape in order to increase the strength and, more importantly, ductility of the pier during major overload situations. Such situations could include earthquake or vehicle impact resistance. This application has been widely adopted in the USA and Japan.

3.3.1 Cleddau Bridge, Milford Haven, 2000

What? The Cleddau bridge, first opened to traffic in 1975, is an 820 m long 7-span continuous steel box girder, which has undergone a structural assessment to the new loading rules and was found to be in need of strengthening. In particular, the tops of the piers had high concentrated forces below the bearings, causing unacceptable bursting stresses in the concrete.

How / what materials? Wrapping of columns with carbon fibre reinforced tape.

Why? The alternative to composite wrapping at the pier top was the design of a steel frame that would contain the bursting forces. This would have required significant temporary framing in a very difficult environment (30 m to 40 m above ground and water) and the final result would have compromised the aesthetics of the bridge with a highly visible steel framework at the top of the six piers. Composite wrapping was therefore seen as the best solution with respect to ease of application, using the existing service gantries, and as a final product, where the composite material would be painted to minimise the visual appearance of the remedial works.

Current condition: The project is just completed, and has achieved the aim of strengthening the bridge without affecting its appearance significantly.

Figure 3.40:
Cleddau Bridge (Concrete Repair Ltd)

Figure 3.41:
Cleddau Bridge installing CFRP (Concrete Repair Ltd)

Figure 3.42:
Cleddau Bridge installing CFRP (Concrete Repair Ltd)

Figure 3.43:
Cleddau Bridge installing CFRP (Concrete Repair Ltd)

3.3.2 Office repair, England, 2000

What? The project is an existing 1980s four-storey concrete framed building with basement. The construction is a braced frame comprising columns, waffle slab and beam floors with stair cores acting as shear cores.

Both the soffit and top surface of the slab have cracked and the cracking is that typically associated with excessive deflection and flexural movement of the slab. Crack widths vary from hairline to in excess of 1 mm around columns. The primary cause of the cracking and movement of the slabs has been shown to be insufficient reinforcement in the column strips of the flat slab and excessive cover.

How / what materials? The existing concrete slabs will be strengthened with externally bonded CFRP laminates. Retrospective analysis has shown that both the hogging capacity and the sagging capacity of the columns strips require enhancement by CFRP laminate strengthening. A total of 8.5 km CFRP laminates is to be bonded to the floor slabs.

The imposed load during plate bonding operations will be restricted. The design criteria to determine the quantity of CFRP laminates will be to limit the strain in the plates while ensuring strain compatibility between the concrete, internal rebar and external plates. Stresses will be re-calculated to demonstrate that they are within acceptable limits. Calculations showed that the strengthening was not required in the event of fire as there would be sufficient residual strength in the slab.

Why? CFRP offered the most cost-effective method of reinforcing the slabs.

Current condition: This work is underway at the time of writing.

Figure 3.44:
The cracked slab (Ove Arup)

Figure 3.45:
Carbon plates being prepared (Ove Arup)

Figure 3.46:
Carbon plates being installed (Ove Arup)

3.3.3 Tickford Bridge, Newport Pagnell, UK, 1999

What? Tickford Bridge is believed to be the oldest operational cast iron highway bridge in the world. It was built in 1810 and is a scheduled ancient monument.

How / what materials? CFRP prepreg sheet was applied in up to 14 layers to the cast iron components. A continuous filament polyester drape veil was installed to provide insulation and avoid any possible galvanic corrosion.

Why? The bridge was deemed to be under-strength, so that reinforcement was required. CFRP sheets were chosen by virtue of their strength, speed of installation (resulting in minimal disruption) and unaffected aesthetics of the strengthened bridge. The scheme was found to be cheaper than traditional alternatives.

Current condition: Good

Maintenance history: Inspections are carried out at two to three year intervals by Milton Keynes Council. (Previous strengthening of the bridge with reinforced concrete and a "crack stitching" technique for the formwork elements had been carried out in 1976 and 1989 respectively).

Further information from: Article in the *New Civil Engineer* magazine, 23rd November 2000, reporting on an award won by the project.

Cost: £150 000

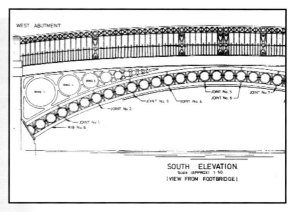

Figure 3.47: Tickford Bridge showing areas reinforced (Milton Keynes City Council)

Figure 3.48:
Tickford Bridge (Maunsell Ltd)

Figure 3.49:
Tickford Bridge, applying reinforcement (Top Bond Plc)

3.3.4 Timber Bridge in Sins, Switzerland, 1992

What? A covered timber bridge in Sins, built in 1807, crosses the Reuss river and consists of two spans, each of about 30 m. Some of the oak crossbeams were deflecting excessively under permissible loading conditions, so it was decided that these crossbeams should be stiffened.

How / what materials? CFRP sheets, with a Young's modulus of 300 GPa, were glued to the underside of the affected crossbeams.

Why? The lightness of the CFRP sheet enabled a quick repair schedule compared with using steel plate. Additionally, the aesthetic qualities of the bridge were hardly affected by the addition of the thin CFRP sheet.

Current condition: Good condition, with the bridge performing according to predictions.

Maintenance history: The quality of the adhesion of sheet to timber was checked by infrared pulse thermography and found to be adequate. A long-term strain measurement programme is underway. To date, no unexpected results have been obtained.

Further information from: Meier U and Winistorfer A (1995)

Figure 3.50:
Sins Bridge (EMPA)

Figure 3.51:
Sins Bridge (EMPA)

3.3.5 Ibach Bridge, Switzerland, 1991

What: Ibach bridge was built in 1969 and became the first concrete bridge in the world to be repaired with CFRP sheet after a steel tendon was accidentally cut. The bridge spans the N2 Highway, as well as the Emme and Reuss rivers. It has an overall length of 228 m and 7 spans.

How / what materials? CFRP sheets, weighing a total of only 6.5 kg, were glued to the soffit of the bridge.

Why? Steel plate would have had a mass of 175 kg, complicating the repair process over a busy highway. The lightness of the CFRP enabled a quick repair schedule.

Current condition: Good condition, with the bridge performing according to predictions.

Maintenance history: A long-term strain measurement programme is underway. To date, no unexpected results have been obtained even under imposed loading tests.

Further information from: Meier U and Winistorfer A (1995)

Ficure 3.52: *Ibach Bridge, general view (Urs Meier)*

Figure 3.53: *Ibach Bridge, plates being installed (Urs Meier)*

3.3.6 Refurbishment of the Cathedral of Christ the King, Liverpool, 1990–2000

What? The Roman Catholic cathedral, which was completed in the 1960s, had concrete stanchions that were clad (Figure 3.54) with mosaic tiles. As a result of debonding between the tiles and the mortar, or between the mortar and the structural concrete, large lumps of material fell from the building in the cathedral precinct, the danger of this meant that recladding became essential.

How / what materials? Glass-fibre reinforced polyester, with a fire-retardant resin, surfaced with a marine-grade gel coat, finished with a granite effect. Overlapping GRP components cover the original damaged cladding and prevent further extension of the damage to the mosaic tile layer.

Why? The choice was between stainless steel and GRP, but GRP was preferred because stainless steel would have required regular cleaning to prevent corrosion in the marine atmosphere of Liverpool. GRP, on the other hand, has proved to be highly successful in marine applications for over 40 years, the standard marine gel coats giving excellent gloss and colour retention. The gel coat plus fire-retardant resin results in a Class 0/1 cladding system.

Current condition: Too recent to show any effect.

Maintenance history: None

Further information from: Norwood L S, 1999, *Glass-reinforced plastic for construction*, Proceedings of a conference on Composites and Plastics in Construction, BRE, Watford, UK (RAPRA Technology, Shawbury, Shrewsbury, UK), paper 3, 1–11

Cost: Not known

Added note: It is of interest to point out that this cathedral also had one of the earliest examples of the artistic use of sculptural GRP panels in buildings, as shown in the third figure.

Figure 3.54: *Close-up view of one of the stanchions that were re-clad with GRP (Bryan Harris)*

Figure 3.55: *Decorative GRP door moulding (Bryan Harris)*

3.4   WATER RELATED PRODUCTS

In the water sector, the driving feature for design it is the ability to survive in the wet and often exposed environment. Steel and other metal solutions need regular maintenance to ensure their long-term servicibility, bringing significant ongoing costs. Therefore the FRP composite alternatives have had considerable success in a variety of applications, some of which are described below.

3.4.1 Drainage grating at Addenbrookes Hospital, Cambridge, UK, 1999

What? Cover over drain allowing rainwater to enter the drainage system. Each unit measures 123 mm by 500 mm, and meets the required standards in load class tests.

How / what materials? Polyester resin prepared as sheet moulding compound. Heat and pressure moulding process utilising a combination of unidirectional glass fibre and chopped strand.

Figure 3.56:
Drainage grating (ACO Technologies)

Why? The dominant reason is the corrosion resistance of the material over cast iron or steel alternatives, particularly in un-trafficked areas where rust build up is greater.

Zero scrap value as compared to cast iron is also a benefit where theft is an issue.

Current condition: Recently installed – still in good condition.

Maintenance history: Similar products and materials installed since 1991 have required zero maintenance and are UV stable.

Cost: Comparable to cast iron equivalent products.

3.4.2 GRP Covers for sewage tanks, Horsham, UK, 1990

What? Two sludge-thickening tanks at Horsham Sewage Treatment Works were covered with a composite roof. Each tank is 10 m in diameter.

Figure 3.57:
Horsham sewage tank covers (Maunsell Ltd)

How / what materials? Pultruded GRP panels were bonded together with epoxy adhesive.

Why? GRP was chosen because of its lightness, impact resistance, strength over a wide flat area, durability and chemical resistance.

Cost: £20 000

3.4.3 Water Tanks: pre-1980

What? Two examples of typical water tank installations in the UK. Many are also installed in the Middle East.

How / what materials? Compression moulding from sheet moulding compound. This provides a cheap automated process for large volumes of similar parts. Some tanks are up to 3 m tall and the panels are typically 1 m square.

Why? Corrosion resistance; low maintenance requirements; cost; UV resistance in the Middle East.

Current condition: There has been a slight dulling of the surface in some cases. Structural performance has remained unchanged.

Maintenance history: Many of the tanks are over 20-years-old, and have had little or no maintenance

Figure 3.58
Sectional SMC water tank (Scott Bader Co Ltd)

Figure 3.59
Sectional SMC water tank (Scott Bader Co Ltd)

3.5 OTHER EXAMPLES

This section gives a number of case studies from a wide range of construction and non-construction sectors, to illustrate the variety of different ways in which FRP composites can be used. It is instructive to look at other sectors where FRP composites dominate, and ask why this is the case. This applies particularly to the boat building industry and the fast-growing wind turbine sector, but there are many applications in aerospace, cars and other vehicles, and the chemical processing sectors where it applies equally.

We can learn from these other applications, and the more novel ideas from construction in order to develop the best ways to use the materials.

3.5.1 Wind Wand, London, UK, 2000

What? A 50 m free-standing glass-fibre-reinforced plastic windwand that is said to be the world's tallest sculpture. The idea was to produce a slender column that would move "interestingly" in light winds.

How / what materials? Epoxy resin reinforced with glass fibre was used to give the correct laminate stiffness. It was then further reinforced with unidirectional carbon-fibre tape. The windwand was manufactured in two halves and then shipped in one piece along the River Thames to Canary Wharf, London.

The windwand weighs approximately 300 kg, has a base diameter of 400 mm decreasing to 80 mm at the top and possesses some 2000 red, high intensity lights at the tip. It is fixed to the ground by 12 bolts, which are anchored into a hinged steel plate that allows the mast to be lowered for maintenance.

Why? Only a polymer composite could give the required complex mixture of stability and flexibility. The design of the wand had to allow for a continuously varying amount of stiffness along its length.

Further information from: *Reinforced Plastics magazine*, July 2000, via www.reinforcedplastic.com.

Notes: Analysis of the windwand has shown that the tip moves as much as 4 m off the upright in 11 mph winds and in storm conditions up to 26 m.

Figure 3.60: *The Wind Wand (Andrew Cripps)*

Figure 3.61: *The Wind Wand (SP Systems Ltd)*

CIRIA C564

3.5.2 Cliff stabilisation, Devon, UK, 2000

What? Rock bolts are being used to hold netting in place to help stabilise the cliff face.

How / what materials? Pultruded glass/polyester rock bolts, grouted into cement.

Why? Sufficient strength is needed to hold the wire netting in place. However it is more important that the rock bolts are lightweight, allowing their installation by "roped up" workers

Further information from: Not available

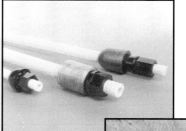

Figure 3.62:
Rock bolts (Weldgrip)

Figure 3.63:
The cliff with bolts installed (Weldgrip)

Figure 3.64
Installation of rock bolts by abseilers (Weldgrip)

Figure 3.65:
The protected cliff (Weldgrip)

3.5.3 Millennium Rest Zone, Greenwich, London, 1999

What? The Rest Zone was one of the 14 pavilions within the Millennium Dome at Greenwich. It fits into a very limited floor plan. The sculptural form was designed as a relaxing, calm environment, in deliberate contrast to the other exhibits.

How / what materials? The structure is a ribbed shell, the shell being of FRP composite, and the ribs being of plywood box beams.

Why? FRP composites were chosen for the shell material, principally to allow the entire structure to be pre-fabricated off site in large panels, and be erected on site in a very short period. The pre-fabricated sections came to site with the box-beams already in place, and the joints were laminated together *in situ*.

Weight was also a significant factor, with the Blackwell Tunnel immediately below this part of the site.

Current condition: Still nearly new

Cost: Not public information

Figure 3.66:
The rest zone (Buro Happold)

3.5.4 MOMI structure, hospitality tent, 1992

Figure 3.67:
MOMI structure during erection (Alistair Lenczner, Ove Arup & Partners)

What? The structure is a response to the brief set by the museum of the moving image (MOMI) for a high quality demountable, hospitality building to be used for a wide variety of functions.

How / what materials? A lightweight, temporary structure, which is constructed from elliptical GRP ribs and a PTFE woven membrane. Its dimensions are 28.8 m x 9.6 m providing a capacity for up to 450 people.

Figure 3.68:
Complete building from outside (Peter Mackinven, Ove Arup & Partners)

Pairs of inclined arches support the fabric; each is constructed from two 32 mm diameter GRP rods, which are connected through a stainless steel spigot and socket joint. It is tensioned into the partial elliptical shape by a series of stainless steel props and cables. The GRP is bonded to the cup of the prop with epoxy glue, which is tightened until the desired curvature achieved. To accommodate the stresses at the base fixing, the GRP rods are connected to mild steel circular hollow sections.

Why? The GRP ribs ensure an elegance and lightness. The structure is minimal and the components once demounted fit into three lorries, while the fabric fits into a box just 1 m³.

Current condition: The structure is currently in storage and not fit for use.

Figure 3.69:
MOMI structure interior view (Ove Arup & Partners)

Maintenance history: The structure was erected and dismantled three times on the South Bank site and also used in another London location adjacent to high rise buildings. On the night of the 8th December 1994 storm force winds destroyed the main membrane. The building could be used again once a new membrane is manufactured.

Further information from: A J working details, (1992) Temporary structure: Hospitality tent, *The Architects' Journal*, Vol.196, No.6, p.34-37

Kronenberg R (1996) *Portable Architecture*, (Architectural Press, Oxford) 30–35

Pawley M (1993) Future systems, *The Story of Tomorrow*, (Phaidon, London)

Cost: Not known

3.5.5 Advanced composite carwash, Hornchurch, UK, 1991

What? The building houses carwash equipment. It is 11 m long, 5 m wide and 4 m high. It is an all-composite structure with no additional support framing.

How / what materials? GRP panels, typically 2.5 m in width and 4 m or 5 m in length, were bonded together with epoxy adhesive in order to produce a membrane structure in which the walls and roof provide the necessary structural integrity. Construction took 1 week to complete.

Figure 3.70:
Hornchurch carwash (Maunsell Ltd)

Why? GRP was chosen because of its inherent durability in this corrosive atmosphere, its high-quality finish and its rapid erection time (which in turn is because it is lightweight).

Current condition: The structure has been removed as a consequence of underground tank replacement.

Cost: £25 000

3.5.6 M25 variable message gantries, London, UK, 1991

What? The department of Transport (DoT) implemented a pilot scheme for variable sign gantries to be used on the M25 in 1991. They are large cantilever structures, carrying signs 2.6 m high, by either 6.4 m or 8.2 m wide.

How / what materials? The cantilevered portion consists of a fully-enclosed GRP box, stiffened internally by a steel lattice, which itself extends from a larger-diameter tubular steel mast. The GRP box therefore protects both the internal electronic equipment within the sign gantry and inspection teams during maintenance.

Why? GRP was chosen because of its lightness, aesthetics and low-maintenance characteristics, in a situation where there are high disruption costs associated with access for maintenance.

Current condition: Good

Maintenance history: No structural maintenance has needed to be carried out.

Cost: £600 000

Figure 3.71: *M25 message gantries (Maunsell Ltd)*

Figure 3.72: *M25 message gantries (Maunsell Ltd)*

3.5.7 Wind Turbines, throughout the world, since 1976

What? A typical aerogenerator, with rotor blades up to 20 m in length, can generate 500 kW at wind speeds up to 25 m/s, rotating at up to 60 rpm. Design life is of the order of 25 years.

How / what materials? Glass-fibre-reinforced polyester GRP (carbon fibres can also be added for extra stiffening, and epoxy resin is also used). Spar/shell construction (as in helicopter blades) with internal webs and foam for stiffening. The load-bearing spar is tape wound and the aerodynamic shell is contact moulded (sometimes press-moulded).

Why? Requirements are smooth output, balanced gyroscopic forces, and low wind-velocity start-up (about 4 m/s). For competitive energy costs, the blades must be made in a cost-effective way and their mass must be effectively used. The materials must possess adequate resistance to fatigue from the cyclic stresses as blades pass the column, must not lose stiffness by excessive moisture uptake, and must have resistance to damage by impact from ice particles and erosion by raindrops in addition to good corrosion resistance. GRP materials meet these requirements.

Current condition: Good

Maintenance history: None

Further information: Mayer R M (Ed) 1996, *Design of composite structures against fatigue: applications to wind turbine blades*, (NMEP, Bury St Edmunds)
Materials Selection Case Study, DTI/EPSRC LINK Structural Composites Programme

Cost: Varies with size

Figure 3.73: *Windmill (SP Systems Ltd)*

Figure 3.74: *Windmill farm in Tarifa, Spain (SP Systems Ltd)*

3.5.8 Hull for Severn class high-speed lifeboat

What? The 17 m Severn class and the smaller 14 m Trent class lifeboats have been introduced by the RLNI to replace the existing Arun and Waverley class boats after 20 years of service.

These boats were intended to have a maximum speed of 25 knots, by comparison with 18 knots for older designs, and a range of 250 nautical miles.

How / what materials? Strength and lightness were achieved by the use of combinations of glass and aramid fibre reinforcement in an epoxy laminating resin: no gel-coat is used.

Figure 3.75: *Severn Class lifeboat of the Royal National Lifeboat Institution (RNLI)*

The hull is manufactured from pre-preg materials in the outer skin and is vacuum-consolidated at a temperature of about 80°C. Sandwich construction with a foamed PVC core is used for the topsides, the inner skin being glass/aramid hybrid mix, which is laid up wet and then vacuum consolidated.

Why? The hull panels were to sustain an increase in loading from 0.31 MPa to 0.48 MPa but with a concomitant saving in weight and improvement in impact resistance that could not be achieved with wood, aluminium or steel. The 55 per cent increase in panel design load specification was achieved with a 12 per cent saving in weight by using the composite design

Further information from: *The RLNI's Severn Class Lifeboat: A Millennium Product*, RLNI, Poole, Dorset, UK

Hudson F D Hicks I A, Cripps R M, 1993, The design and development of modern lifeboats, *J Power & Energy* (Proc I Mech E; A), 207, 3–12

Cost: £1.8 million

4   Applications with potential for growth

4.1   INTRODUCTION

What is clear from the case studies of existing projects involving FRP composites is that they are most successful when they work with one or more of the positive features of the materials. They are less successful when they try to be something else – simple substitution rarely works. Therefore successful future applications will be those that build on the strengths of the materials, are designed in a holistic way based on the properties of FRP composites and meet a real need from the industry.

It is worth noting that polymers currently use less than 5 per cent of the by-products of oil refining, so there is scope for increased use, without the use of additional crude oil.

This chapter seeks to introduce a range of ideas that can be developed further in the market place. Most of them already exist in small numbers or as prototypes and are therefore in need of development and appropriate marketing, rather than fundamental research.

As discussed before, the main positive features that can be provided by FRP composites for new construction are:

- high strength-to-weight ratio
- resistance to weathering / harsh environments. Both "cold and harsh" and "hot and harsh" (hot weather and hot processes) can be dealt with
- impact resistance
- potential for low imflammability
- some are radio transparent and non-magnetic
- many provide flexibility of appearance
- potential for rapid installation
- low maintenance requirements.

Remember that the actual properties of a particular FRP composite will represent a balance between these. No single FRP composite has all of these properties.

4.2   AREAS FOR GROWTH

The following have all been suggested as deserving further development as potential applications for FRP composites in the construction sector.

4.2.1   Building/architectural

FRP composites may be considered for a variety of construction applications where long-term durability and minimum maintenance are required combined with their mechanical properties of high-strength and low-weight. The ability to form complex moulded structures in a variety of coloured finishes also provides a greater degree of architectural freedom.

Building/Architectural		Section/figure
Application	**Notes**	
System/prefabricated buildings	eg for Schools Military Hospitals Hotels Houses	Figure 4.7 Figure 4.10
Mobile/demountable structures	Shops Temporary car parking Emergency shelters Military Leisure/stadia/hospitality	Section 4.3.3
Monocoque structures/complete shell structures	Variety of building types	Section 3.1.1 Figures 4.2 and 4.9
Products and components	Bathroom components Stairs Roofs Chimneys Handrails Columns Balconies Doors and roller doors Snap joints Window and other frames	Figure 4.6 Figure 4.13 Figure 4.12 Figure 2.3 Figure 4.1 Figure 4.38 Figure 4.8
Moulded integral features	Restoration purposes	Section 3.3.6 and figures 4.4, 4.12
Timber strengthening	Allows customisation of tensile, compressive or shear strength	Section 3.3.4
Air conditioning components	Able to resist the environment involved with processing warm, sometimes moist and polluted air – these are already used but could be expanded further	
Permanent formwork/exoskeletal reinforcing	Can replace internal reinforcement	

Figure 4.1: *Hotel balcony (Strongwell)*

Figure 4.2: *Dome (Composites Janero, Spain*

Figure 4.3: *Dormer Windows (Composites Janero, Spain)*

Figure 4.4: Stone façade including repairs in FRP composite (Composites Janero, Spain)

Figure 4.5: *"Bulls eye" window (Composites Janero, Spain)*

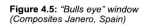

Figure 4.6: A bath (Scott Bader Co Ltd)

Figure 4.7: Gatehouse cabin (Scott Bader Co Ltd)

Figure 4.8: *Rooflight (Composites Janero, Spain)*

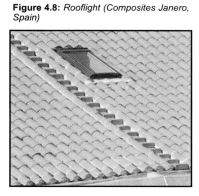

Figure 4.9: Mosque Domes (CETEC Consultancy Ltd)

Figure 4.10 Tram shelter, Blackpool (Scott Bader Co Ltd)

Figure 4.11: *Chimney Unit (Readings Composites Ltd)*

Figure 4.12: Chimney unit in factory (Readings Composites Ltd)

Figure 4.13: Octagonal roof (Scott Bader Co Ltd)

Figure 4.14: Manchester City stadium roof (Scott Bader Co Ltd)

4.2.2 Bridges

The ability to tailor FRP composites to enable high-strength with low-weight, while achieving resistance to corrosion under most environmental conditions, makes them a favourable option for bridge applications. They are increasingly being used to construct new bridges as well as provide repairs to existing structures. New bridge construction offers scope for the further development of standard pultruded and modular sections using a high quality continuous manufacturing process.

Bridges		
Application	**Notes**	**Section/figure**
Footbridges	Structure and post-tensioning stay cables	Sections 3.2.1, 3.2.2, 3.2.3, 3.2.5, 3.2.6, 3.2.4, Figure 4.17
Lifting road bridges	Clear weight benefit for the lifting part	Section 3.2.4
Replacement decks	Reduce time and install cost	Figure 4.18
Enclosures and strengthening	Resist harsh environment, protect existing structure	Section 3.3.6
Long span bridges	Potential to reduce support structure	Section 3.2.6
Bailey Bridges	Weight is key feature – some examples of this exist already	
Modular bridges for disaster relief	Low weight to allow air freight/ helicopter installation	
Components such as gantries, bearings and expansion joints	In steel, require substantial maintenance and are prone to fatigue damage	Figures 4.15 and 4.16
Floating pontoons	Could be collapsible	

Figure 4.15: *Forth bridge gantry (Composite Solutions Ltd)*

Figure 4.16: *Forth bridge gantry (Composite Solutions Ltd)*

Figure 4.17: *168 m long walkway (Composite Solutions Ltd)*

Figure 4.18: *Toms Bridge (Strongwell)*

4.2.3   Repair and maintenance

FRP composites can be used for repair, seismic retrofitting and upgading to extend the life of existing structures.

Repair and maintenance		
Application	**Notes**	**Section/figure**
Retrofit column wrapping	Saves time and scaffolding – already in the market but more can be done, adds ducility and strength	Section 3.3.1
Soffit/ceiling strengthening		Section 3.3.2
Timber strengthening	Allows customisation of tensile compressive or shear strength	Section 3.3.4
Brickwork strengthening	Especially for seismic loads	
Metal strengthening		Section 3.3.3

4.2.4   Marine and geotechnical

The non-corrosive properties of FRP composites combined with their ability to withstand harsh environments make them particularly suitable for marine applications.

Marine/geotechnical		
Application	**Notes**	**Case studies/pictures**
Harbours	Pier decks and gravity fenders	Figures 4.21 and 4.22
Piling		Figures 4.19 and 4.20
Canal works		
River training		
Rock bolts	Used for securing netting to rocks, for temporary supports and to resist ground slippage	Section 3.5.2

Figure 4.19:
Sheet piling installation (Creative Pultrusions Inc)

Figure 4.20:
Sheet piling in place (Creative Pultrusions Inc)

Figure 4.21:
Gravity fender in Qatar prior to installation (Eurocrete Ltd)

Figure 4.22:
Gravity fender in Qatar in place (Eurocrete Ltd)

4.2.5 Water industry

The corrosion resistant qualities of FRP composites make them suitable for applications that involve contact with water.

Water industry		
Application	**Notes**	**Case studies/pictures**
Tanks	Already done but large market	Section 3.4.3
Roofs	Need light weight and durability	Section 3.4.2
Access system	eg drain covers	Section 3.4.1
Swimming pool	Exist overseas – may become popular in UK	Figures 4.23 and 4.24

Figure 4.23:
The swimming pool being installed (Poolquip Nederland BV)

Figure 4.24:
The swimming pool installed (Poolquip Nederland BV)

4.2.6 Other

High performance FRP composites can now be considered for a diverse range of applications across construction and beyond.

Other		
Application	**Notes**	**Case studies/pictures**
Wind powered turbines	Strength-to-weight and fatigue important – already dominate but a fast growing market	Section 3.5.7
Masts eg telephone	Strength to weight important. Can use "snap joints" – used in Australia and USA	Figure 4.25
Structural frames	Strength to weight important	Figure 4.26
Chimneys	Ideal because of strong resistance to harsh environment – already wide use in chemical industry	Figures 4.11 and 4.12
Cooling towers	Ideal because of strong resistance to harsh environment	
Smart components	Opportunities to incorporate sensors and adaptive behaviour in the materials	
Rail fixings	Prone to fatigue damage in steel, so ideal in FRP. Used in USA	
Heavy duty Power Towers	In remote areas, where construction would be easier (using helicopters) and maintenance and inspection could be minimised	
Crash barriers		
Scaffolding		
Re-usable formwork		
Architectural/sculptural		Figures 4.27, 4.28, 4.29 and 4.30

Figure 4.25: *Power transmission pole (Strongwell)*

Figure 4.26: *Large scale frame (CETEC Consultancy Ltd)*

Figure 4.27: *Sculpture in London (Andrew Cripps)*

Figure 4.28: *Fabric structures (Ann Alderson)*

Figure 4.29: *Giant "Lego" block (Readings Composites Ltd)*

Figure 4.30: *Statue cover (Readings Composites Ltd)*

4.3   DEVELOPED IDEAS FOR SELECTED APPLICATIONS

This section reviews a number of products or prototypes that have been developed either in the UK or abroad, and show potential for great expansion in the future.

4.3.1   Blast panels

Blast resistant cladding panels were developed and tested as part of a DTI Carrier Technology Programme project (Alderson 1999). The FRP composite solution provided the required blast resistance and yet was light enough not to incur other building penalties in terms of additional structure. The FRP composite also allows a wide choice of surface appearance, reducing the evidence of fortification.

Figure 4.31: *The Panel (Richard Walker, Pera)*

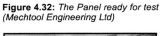

Figure 4.32: *The Panel ready for test (Mechtool Engineering Ltd)*

Figure 4.33: *The explosive test (Mechtool Engineering Ltd)*

Cladding panels comprise a lightweight core of polyurethane and foamed clay prills sandwiched between multi-axial glass fibre skins, both variable in construction to provide a match for required strength and durability. Panels are available in a range of colours and surface finishes. Tests have shown that with appropriate design they can meet the most stringent blast performance.

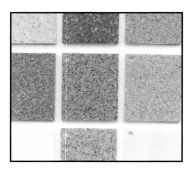

Figure 4.34: *Example panel finishes (Mechtool Engineering Ltd)*

4.3.2 Rain shelter for railway stations

It is often noted that the solutions that work best with FRP composites will be those where the materials are working both as structure and panel. This project has demonstrated the possibility of using FRP composites to produce a practical rain shelter for use in railway stations that is quick to install.

The project developed a single-piece, lightweight railway station canopy as shown in the figures below. The asymmetrical shape helps with the use of space on the platform, with the longer side (3 m) over the edge of the platform, and the shorter side (1.2 m) in the centre. The column to column distance is 10 m, allowing the minimum number of columns, assisting the speed of installation.

In terms of materials, the horizontal beam is a square section carbon/glass fibre box beam. This weighs around half that of the equivalent steel structure. The canopy consists of a low-cost core material covered in a glass fibre composite to give sufficient stiffness.

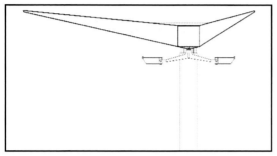

Figure 4.36: *Prototype of canopy (Brooks, Stacey, Randall)*

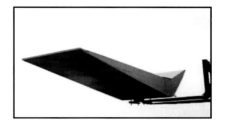

Figure 4.35: *Section of canopy (Brooks Stacey Randall)*

For further information refer to Stacey (1999).

4.3.3 Prototype military shelter

Pultruded glass fibre sections have been used in the primary structure for a military shelter prototype. The trusses used a purpose-designed section for the primary load bearing member. This is bent elastically into an arch that is restrained with a web of bias-cut sailcloth fabric. The ribs in the truss have a section 92 mm wide by 13 mm deep. This allows the rib to be bent to a radius of 1.5 m allowing for adequate capacity to deal with additional stresses due to external loading. The sections have also been designed to simplify fabrication and assembly by incorporating jigging slots.

Figure 4.37: *Test structure (Lightweight Structures Unit, Dundee University)*

Figure 4.38: *Clip-joint design (Lightweight Structures Unit, Dundee University)*

4.3.4 Pultruded GRP rods bonded into wood

Timber is widely regarded as a very "green" building material, since it is produced from a renewable source. However it is not always able to perform well enough to meet the demands of construction projects. Most often the problem occurs with transferring loads at junctions. One possible solution to this, in some circumstances, is to use rods bonded into the wood to help with these junctions, and transfer the load more efficiently. A number of systems have been investigated, including both steel and pultruded GRP. One particular benefit of the GRP solution is that it does not risk damage to saw blades if the timber is cut later on in life during recycling. For further information refer to Harvey *et al*, 2000.

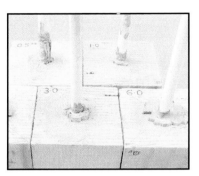

Figure 4.39: *Samples after testing to failure (Martin Ansell, Bath University)*

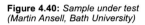

Figure 4.40: *Sample under test (Martin Ansell, Bath University)*

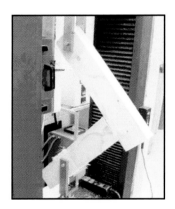

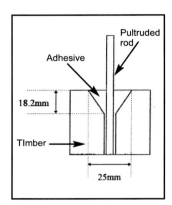

Figure 4.41: *Schematic of counter-bored wood with bonded in rod (Martin Ansell, Bath University)*

5 Designing with fibre-reinforced polymer composites

This section is aimed particularly at engineers and specifiers who are relatively unfamiliar with FRP composites, and who would not normally have considered them for the application they are in the process of designing, whether structural or non-structural. The scope and level of this chapter is, therefore, not meant to replace existing publications giving detailed design guidance, but to equip and raise the awareness of designers to the main factors that should be considered in determining whether a FRP composite will successfully fulfil the main design criteria.

The first part of this chapter discusses how the choice of material might be made. The remaining parts describe design approaches to using the chosen FRP composite, give limited design information to start the process, and references for further information. It is important to emphasise that manufacturers' own product data must be used for detailed design and manufacturers/suppliers must be consulted at an early stage, otherwise, extensive material testing will be required.

5.1 MATERIALS SELECTION

In most applications where FRP composites are used there are alternative available materials. For example (to take a non-construction example), if one were building a 15 m to 20 m boat it could be made out of the following materials:

- traditional timber construction
- timber/epoxy strip planking
- plywood
- steel or aluminium
- ferro-cement
- GRP or CFRP.

The choice would depend on the skills, facilities and materials available and, of course, the cost. The form of the boat could be identical but there might be other considerations, such as performance in use where weight may be important, or maintenance and durability. These considerations affect the choice of materials for all components in construction. For FRP composites, another factor would be the number of items of the particular design that the designer is planning to produce, as this would affect the choice of manufacturing process. For many kinds of boats GRP has become the material of choice because it provides the best fit to the requirements at the most competitive price.

Precisely the same range of issues are valid with construction examples, and depending on circumstances, FRP composites can be the best choice.

5.1.1 Key factors in choosing an appropriate material

This section covers factors that play a role in deciding which material is appropriate for the application being considered, and is not just about FRP composites. One of them may be the overriding factor in the design team's criteria, but each one needs to be considered. In each case the issues relevant for FRP composites are discussed briefly, as it is assumed that the reader will know about the alternative materials.

Given the great variety of applications for which FRP composites can be used, it is difficult to be more specific about what should be chosen at a particular time. Therefore, this section sets out the issues to consider, when comparing FRP composites with alternative material choices.

There are a large number of different sources of far more detailed information on the selection of materials for engineers. Amongst the best known are those by Ashby (1999), embodied in the *Cambridge Materials Selector*, and by Charles, Crane and Furness (1997).

Material properties

It is clear that the choice of a material depends on the basic properties of strength, stiffness, weight etc. There may be overriding benefits to be gained from the selection of FRP composites in terms of stiffness/weight. Generally there will be a choice of materials to achieve the required properties, and then other features will also need to be considered. These issues are developed further in sections 5.1.2 and 5.2.1, and there are representative tables of data.

Form/3D elements

The form of the component under consideration will be defined by the functional requirements or by the particular aesthetic that the designer is aiming for. Generally speaking the form will not be dictated by the material, although some materials will give a greater range of choice. FRP composites lend themselves well to the forming of complex shapes.

Table 5.1:   *Main characteristics of the common manufacturing processes for FRP composites*

Process	Limitations: shape	Limitations: dimensions	Moulded surface finish	Production volumes	Production rate parts/mould /day	Labour content
Hand lay-up	no limitations	no limits	1 face only	up to 100	1 to 4	high but unskilled
Spray-up	no limitations	minimum of 0.25 m × 0.25 m	1 face only	up to 100		high and semi-skilled
RTM/Low pressure press moulding	must be demouldable ie no undercuts without complex tooling	1–5 sq. m	both faces	100 to 2000	2 to 30	not high, but semi-skilled
Hot press moulding and DMC/SMC	complex shapes possible within limitations of demouldability	1–5 sq. m Possibility of multi-cavity moulds	both faces	over 5000	80 to 700	low
Filament winding	geodesic	practically unlimited	interior smooth face	up to 2000	1 to 8	small but skilled
Pultrusion	constant section profiles	unlimited length, section of order 1 m × 1 m hollow	all faces smooth			small but skilled
Continuous sheet production	flat sheet with constant cross section	limited by machine width (usually 1 m)	both faces	300–1000 kg/ hour	3–10 m/min	low

Method of manufacture

All materials have their own manufacturing methods; the choice is influenced by the availability of the necessary time, facilities and skills. Within the field of FRP composites, there are many available manufacturing processes and these will influence the cost of the units as well as the quality of finish, shape and structural performance, etc. The principal methods of manufacture and their implications for the product are discussed in Chapter 2. An important factor in selecting the method of manufacture is the number of repeat units required. Table 5.1 gives some information, which helps to inform the choice of manufacturing process.

Programme

Different materials with different production methods have differing times for design development, shop drawings, materials purchase, and fabrication. The same applies to the manufacturing methods used for FRP composites. While Chapter 2 gives a guide to these, each design will be different and will have to be investigated on its own merits.

Cost

Cost is always an important consideration in the selection of the material and manufacturing process. Contractors' costs can be broadly broken down into the following categories:

- overheads, profit and risk
- engineering, drawing office and management
- raw materials
- tooling
- labour and plant for manufacture
- transport and avoidance of disruption to traffic
- installation on site, handling
- commissioning and maintenance.

These should be included in making a fair comparison between solutions. Designers should also consider the whole-life costs of a particular product so that the costs of maintenance, repair and eventual disposal are included. This will generally be more supportive of the use of FRP composites than the initial cost alone.

Fire

In order to determine the main design criteria, with respect to fire, one must consider whether it needs to have a certain fire resistance, limit the spread of flames, or avoid the generation of smoke or other by-products of fire. In some instances none of these apply, but at other times these requirements can be very stringent.

From the FRP composite perspective, the next issue is whether the fire related needs can be met and how this can be done. These issues are discussed further in Section 5.2.3.

Environmental effects on the material

Clearly the intended application for the product is of great importance to the choice of material, as well as the environment in which it will be placed, and any natural or chemical agents acting on it. Clearly the most common of these are water, hot/cold temperature cycling and UV radiation. There may also be air pollutants, and salt spray from the sea or from road treatments can be important in some locations.

FRP composite materials generally perform well in harsh environments, but the choice of materials and surface finishes is important to optimise performance. This can be seen from their widespread use in the marine, chemical and other processing environments. These issues are discussed further in Section 5.2.2.

Re-cycling/sustainability

The issues of recycling and sustainability continue to grow in importance in public and political spheres. Therefore the designer must consider these issues (further discussion is given in Section 5.2.5). The first issue to be considered is the life-cycle of the application, ie how long is it expected to be needed for, and is the solution being considered appropriate to that time scale – often solutions are over-specified in all materials. However, flexibility is becoming increasingly important in sustainable design, since it allows for the future adaptation of the building or other structure, hence maximising its useful life.

The end-of-use properties of a material are also important in some circumstances, and the issues of re-use, recycling or safe disposal of materials should be considered at the design stage. Many recyclable materials are not in fact recycled, as a consequence of the difficulties of separation and transport, and the tendency of material properties to degrade with each cycle of the process. However, FRP composite materials are particularly difficult to recycle (Section 5.2.5).

Site works

For some projects there will be site factors or conditions that will significantly affect the material choice. These might include restricted or difficult access, when weight becomes critical, expensive or disruptive locations where time is crucial or sensitive locations where avoiding damage to the existing structures or wildlife are most important. All of these may benefit or restrict the use of FRP composites, but they are often beneficial because of strength to weight issues, and their ability to repair existing structures.

5.1.2 Description of the properties of FRP composites

There are three main types of fibres in common use and three classes of resins. A brief description of each is given below, followed by graphs showing a limited range of properties. Further information is given on other properties in the next main section on designing with composites; this information is intended to help with understanding the choices to be made.

Glass fibres

Glass fibres are used for the majority of composite applications because they are cheaper than the main alternatives. There are different forms known by names like E-glass (the most frequently used), and S2- or R-glass. The main characteristics of glass fibres are their high strengths, moderate Elastic (or Young's) moduli and density, and their low thermal conductivity. Special corrosion resistant glass fibres are also available.

Aramid fibres

Aramid fibres are polymeric fibres, the main characteristics of which are their high strengths, impact resistance due to their energy absorbing properites, moderate Young's moduli and low densities. Laminates formed from aramid fibres are known for their low compressive and shear strengths. The fibres themselves are susceptible to degradation from UV light and moisture but exhibit resistance to acids and alkalis. They are more resistant than other fibres to chemical attack from hydrochloric acid. Kevlar is the trade name of the original aramid patented by Dupont, and Twaron is the name of a similar fibre produced by Hexcel.

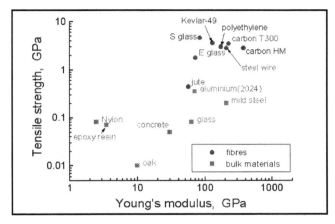

Figure 5.1: *Tensile strength and elastic modulus of some common reinforcing fibres and bulk engineering materials (Bryan Harris)*

Carbon fibres

Carbon fibres, manufactured by the controlled carbonization of organic precursors (such as the textile fibre PAN and pitch), are produced in many grades. The main characteristics of carbon fibres are their high strengths and Young's moduli, and their very low densities and thermal expansivities. The characteristics of these fibres are often indicated in their commercial names by codes such as HS (High Strength), HM (High Modulus), UHM (Ultra High Modulus) etc. The wide range of fibres and properties that are available, offer the maximum possibility for optimisation of the material to provide properties specifically matched to a particular application.

An indication of main strength and stiffness characteristics of these and other fibres, compared with the properties of more familiar bulk materials is shown in Figure 5.1. The strongest, stiffest materials are on the top right of the diagram. This shows that the fibres used in polymer composites are all approximately as strong as steel wire. Carbon fibres can be slightly stiffer than steel, while the other fibres are less stiff. Their initial high-cost can be traded for extremely high performance.

Unsaturated polyester resins

Unsaturated polyester resins are used for the majority of composite structures. They consist of polyesters of relatively low molecular weight, dissolved in styrene. Curing occurs by the polymerisation of the styrene, which forms cross-links across unsaturated sites in the polyester.

Polyester resins are relatively inexpensive, easy to process, allow room temperature cure and have a good balance of mechanical properties and environmental/chemical resistance.

The main issues relating to the use of polyester resins are:

- moderate adhesive properties
- styrene vapour release during cure
- curing is strongly exothermic and can cause damage if processing rates are too high
- shrinkage on cure of up to 8 per cent, although this can be significantly reduced by the use of shrinkage-control additives.

The vinyl esters are a subset of the polyester family, characterised by higher thermal stability (use temperatures up to 125°C, by comparison with about 70°C) and better environmental resistance.

Epoxy resins

Epoxy resins are used for the majority of high-performance composite structures. They are generally two-part systems consisting of an epoxy resin and a hardener which is either an amine or anhydride. A wide variety of formulations are available giving a broad spectrum of properties. The higher-performance epoxies require the application of heat during a controlled curing cycle to achieve the best properties.

Epoxies have excellent environmental and chemical resistance and superior resistance to "hot-wet" conditions. Compared to polyesters, epoxies require more careful processing and are more expensive by a factor of 1.5 to 3. However, epoxies have better mechanical properties, give better performance at elevated temperature and exhibit a much lower degree of shrinkage (2 to 3 per cent). Their use incurs less waste and permits faster production rates, they can therefore be competitive with polyester in terms of cost. Epoxies also give fewer problems with respect to volatiles during processing, although they can cause dermatitis unless carefully handled.

The relatively high-cost of epoxy resins compared to polyester meant that their use was long restricted to the aerospace and leisure industries and as structural adhesives. Only recently have they begun to be used to any great extent in composites for construction. Their high performance, easy processing and strong adhesion have led to an increase in their market share.

Phenolic resins

Phenolic resins are of particular interest in structural applications owing to their flame-retardant properties, low smoke generation, dimensional stability at elevated temperatures (up to 300ºC), and excellent resistance to environmental degradation. Phenolic resins are produced by a condensation process from reacting phenol with formaldehyde.

Phenolics exhibit good dimensional stability and resistance to acids. Undesirable features of phenolics are their relatively low toughness and the generation of water during curing. The latter point is important, because a phenolic that is not fully cured will give off water in the form of steam during a fire, which can cause failure of the laminate. They are also limited in appearance due to their dark colouring.

5.2 DESIGNING WITH FRP COMPOSITES

Throughout this section different aspects of designing with FRP composites are explored in more detail. The intention is to give the reader a strong position to start with a design, but it is by no means a definitive guide. The aspects addressed are:

- structural design criteria
- fire
- environmental issues
- durability
- finishes and aesthetics.

5.2.1 Structural design criteria

Structural design in FRP composites

An engineer familiar with the simple stress-analysis approach used in most design work will find that this approach is often difficult to apply in the use of composites. There is no equivalent to the steel industry's "blue book", which lists the properties and allowable stresses for a range of standard sections. Equally there are no prescriptive codes of practice that cover the full range of composites. Before embarking on a stress analysis, the designer must choose from an initially bewildering array of materials for both fibre and resin, the orientation of the fibres, and the literally infinite range of section sizes. Note that all materials properties presented in this report are typical average values and supplier data must be used for final calculations.

This section sets out to explain the factors influencing the strength and stiffness of a composite element, enabling the designer to choose appropriate materials and understand how the element will behave under load. When analysing a composite element, a traditional hand calculation is normally a good starting point, although for most structural elements a detailed finite-element analysis is required. This guide assumes that professionals experienced with composite materials will carry out the detailed design.

Selection of materials

As discussed in Section 5.1, purely structural issues such as strength and stiffness are only two factors that influence the choice of materials. Cost, weight, fire resistance and method of construction are other significant factors that cannot be overlooked. Tables 5.2 and 5.3 set out typical properties for the most common fibres and resins, and demonstrate clearly how strength and stiffness are bound with other factors such as cost and weight, (Bunsell 1988, Donnet and Bansal 1990, Yang 1993, Hill *et al* 1999).

Table 5.2: *Ranges of basic data for common fibres*

Fibre	Tensile strength (GPa)	Young's Modulus (GPa)	Density (10^3 kg/m³)	Cost (£/kg)
Aramid	3.15–3.60	58–130	1.39–1.47	20
Carbon	2.10–5.5	200–500	1.74–2.20	10–200
Glass	2.4–3.5	72–87	2.46–2.58	2.5

Table 5.2 reinforces both the range of choices available to the designer and the issues that must be considered when making a choice. It also starts to distinguish between the characteristics of the fibres. For example the relative cheapness of glass fibres is balanced by higher weight, while carbon's attractively high Young's modulus is offset by a higher cost.

Table 5.3: *Ranges of basic data for common resins*

Resin	Tensile strength (MPa)	Young's Modulus (GPa)	Density (10^3 kg/m^2)	Cost (£/kg)
Polyester	50–75	3.1–4.6	1.11–1.25	~ 2.5
Epoxy	60–85	2.6–3.8	1.11–1.20	~ 5–10
Phenolic	60–80	3.0–4.0	1.00–1.25	~ 2

Having explained that material choice is rarely based solely on structural attributes, this section will be restricted to considering composites from a purely structural viewpoint. In simple terms, the principal contributor to the strength and stiffness of a composite is the fibre. The resin has significantly lower strength and stiffness than the fibres, and acts to bind the fibres together, transmit loads into and out of the fibres, and to protect the fibres from mechanical and environmental damage.

Tensile stiffness and the strength of simple FRP composites

The tensile stiffness and strength of a composite element is the most straightforward of the properties that a designer will need to consider. The transfer of fibre properties into the composite needs to be understood by the designer. A full treatment of the theory of composite laminates is beyond the scope of this report, readers are referred to standard texts, such as Jones (1975), Matthews and Rawlings (1994), Powell (1994), and Gibson (1994). However, it is appropriate to give some basic insights into the calculation of mechanical properties.

Stiffness

The tensile strength/weight ratios of FRP composites are high in comparison to steel but their stiffnesses are generally comparatively low. Thus stiffness requirements tend to dominate structural design. As the stiffness of an FRP composite element is dominated by the fibre, the most obvious way to create a stiffer composite is to select a fibre with a higher elastic modulus but a significant cost penalty is usually incurred in doing so. It is therefore recommended to examine various other means of stiffening a composite element such as increasing the thickness of the component, incorporating local variations in thickness or integral ribs, or introducing double curvature.

In a rod of composite in which the fibres are all continuous and arranged along the length of the rod, the fibres and matrix will support load in proportion to the cross section they occupy. Hence we can estimate the Young's modulus, E, with an equation based on simple beam-theory result:

$$E_c = E_f V_f + E_m (1 - V_f) \tag{1}$$

where the subscripts c, f and m refer to composite, matrix and fibres and V_f is the proportion of the cross-section that is occupied by the reinforcement (ie the fibre volume fraction). This expression is known as the "rule of mixtures". It can be seen from tables 5.2 and 5.3 that the fibre/matrix modulus ratio, E_f/E_m varies from about 20 to as high as 300, depending on the fibre type. Hence for most composites with more than about 50 per cent volume of fibres, the stiffness in tension is dominated by the fibres, and the resin contribution can be ignored.

Fibre contents as high as 70 per cent by volume may be achieved in mechanised processes such as filament winding and pultrusion, whereas manual lay-ups and random fibre arrangements produced from materials such as SMC and DMC may produce fibre contents as low as 10 to 25 per cent. Hence this factor is important in estimating the properties of an FRP composite.

A point of major importance in comparing FRP composites with conventional metallic materials is that their densities are significantly lower. Thus in any design where the appropriate material index is not simply stiffness, but one of the modulus/density (E/ρ) ratios, FRP composites demonstrate major advantages, as illustrated in the charts given by Ashby (1999) and the case studies given by Ashby and Jones (1980). A simple comparison of the data in Table 5.4 (adapted from Table 5.2) illustrates the point. For conventional minimum-weight design, the material index is E/ρ, and for such applications only the higher-performing carbon and aramid fibres exhibit advantages. However, for applications involving bending (cantilevers or columns loaded in compression, the appropriate index is E/ρ^2, while for applications involving plate buckling (shells, for example) the index is E/ρ^3. The density advantage increasingly leaves structural steel far behind, although aluminium still competes satisfactorily with GRP.

Table 5.4: *A comparison of values of the combined stiffness/density indices used in structural design for typical unidirectional composites and for steel and aluminium*

Reinforcement type	E/ρ	E/ρ^2	E/ρ^3
Carbon P755	188	108	58
Carbon HM	153	94	58
Carbon T300	93	57	35
Aramid (Kevlar-49)	60	42	29
S2 glass	27	13	6
E glass	23	6	5
Aluminium (5083 alloy; N8)	26	10	4
Steel (grade 50, high-yield)	27	3	0.4

Strength

The strength of a unidirectional composite is more difficult to calculate precisely because it depends strongly on the failure mode, which is often complex in fibre composites, and on the statistical distribution of strengths of individual filaments in a fibre bundle (for example, see Harris (1999) Chapter 4). However, a preliminary estimate can be obtained from a rule-of-mixtures equation similar to equation (1):

$$\sigma_c = \sigma_f V_f + \sigma_m'(1 - V_f) \tag{2}$$

The term f is the notional average strength of the fibres and $'m$ is the stress in the matrix at the point of failure of the fibres, which in polymer-matrix composites, will usually be well below the normal matrix failure strain. Again, given the differences in strength of the common fibres and matrices, it is the strength of the fibres, which dominates the behaviour of the composite.

Equations 1 and 2 allow an estimate of the strength of an FRP composite to be made, based upon average data. Remember that for detailed work, better data will be available from suppliers, and this must then be used. An indication of the strengths and stiffnesses of some unidirectional materials, together with typical properties of structural steel and aluminium, is given in Table 5.5.

Table 5.5: *Estimated properties of several types of unidirectional FRP composites compared to structural steel and aluminium. A fibre volume of 0.65 is assumed for the FRP composites*

Reinforcement type	Typical density (10^3 kg/m³)	Tensile strength (MPa)	Tensile modulus (GPa)
E-glass	2.15	1585	49
S2-glass	2.10	2370	57
Kevlar-49	1.43	2365	86
carbon T300	1.63	2300	151
carbon HM	1.63	1845	248
carbon P755	1.79	1390	337
Aluminium (5083 alloy; N8)	2.7	275–320	70
Steel (grade 50, high-yield)	7.8	up to 620	207

HM is a high-modulus PAN-based carbon fibre

T300 is an intermediate-strength PAN-based carbon fibre

P755 is a high-modulus pitch-based carbon fibre

The properties of unidirectional composites normal to the fibres are dominated by the resin and the fibre/matrix interface. The strength and stiffness in this direction may be one or two orders of magnitude lower than their values in the fibre direction. This high degree of anisotropy, like that of timber, means that designers must exercise great caution in avoiding stress systems that result in significant loads normal to the fibres. Moreover, when the fibres are short, as in moulding compounds, rather than continuous, their reinforcing efficiency is reduced because stress must be transferred into them by way of the matrix and the shear strengths of the matrix, and the fibre/matrix interface are both quite low.

Effect of fibre orientation on strength and stiffness of FRP composite materials

The mechanical properties of the fibres and matrix making up a composite are obviously the key elements influencing the strength and stiffness of the component. However, there are numerous other factors that have a significant influence on both the strength and stiffness of a composite element. The most important of these is the orientation of the fibres. This section gives quite detailed information on how to estimate the effect of combinations of fibres in different directions on the overall capacity of the FRP composite. Some may choose to pass over the theory, but should remember that the maximum capacity is an absolute maximum – other combinations of materials will not achieve the best possible performance if the alignments or materials are not the same.

As equations 1 and 2 show, a composite acts most efficiently when the applied forces are directed along the line of the fibres. If the fibres are at an angle to the applied stress, as some proportion of them will be in most practical composites, their load-carrying capacity is reduced. The properties of a unidirectional FRP composite in the direction transverse to the fibre direction may be an order of magnitude lower than the properties parallel with the fibres. The variation of the elastic modulus of a unidirectional composite as the angle of loading, θ, to the fibre direction varies from 0 to 90° is given by the expression from anisotropic elasticity theory:

$$\frac{1}{E(\theta)} = \frac{\cos^4\theta}{E_0} + \frac{\sin^4\theta}{E_{90}} + \left(\frac{1}{G} - \frac{2v}{E_0}\right)\cos^2\theta \, \sin^2\theta \tag{3}$$

where E_0 and E_{90} are the Young's moduli parallel with and transverse to the fibres, respectively, and v and G are the major Poisson ratio and the in-plane shear modulus.

Although precise calculations of the anisotropic properties of practical composites require the application of the theory of anisotropic elasticity referred to in the texts cited earlier, reasonable estimates can often be made by simpler means. For example, Krenchel (1964) used a simple expression, which sums the contributions from each group of fibres, α_m, lying at a specific angle, θ, to the applied stress. If we assume that the fibres are continuous and that there are no elastic Poisson constraints ($v_f = v_m$), each group of fibres can be considered to have a reinforcing efficiency $\alpha_n\cos^4\theta$, and an overall composite efficiency factor, η_θ, can be defined as:

$$\eta_\theta = \sum a_n \cos^4\theta \tag{4}$$

The approximate composite modulus can then be obtained from a modified version of the rule of mixtures, equation (1) as:

$$E_c = \eta_\theta E_f V_f + E_m(1 - V_f) \tag{5}$$

This is a simplistic approach that gives reasonable results for the tension and bending capacities, and stiffness of section. However, it does not consider the complex interaction between the fibre and the resin experienced by a composite element in compression or shear, and it ignores a possible contribution from fibres normal to the applied stress. For a simple cross-plied glass-fibre/epoxy laminate of overall $V_f = 0.6$, and with lay-up $(0,90)_{3s}$ the Krenchel model, with $\eta_\theta = 0.5$, would predict a Young's modulus in the two orthogonal directions of

0.5 x 70 x 0.6 + 3 x (1– 0.6), or 22.2 GPa, which is reasonably close to experimental values.

Estimations of properties of a multi-laminate FRP composite

The theory discussed above relates to the properties of single layers of FRP composite. For more complex laminates involving a number of layers with different properties, a rule known as the 10 per cent rule (Hart-Smith, 1993) is both simple and reasonably accurate. It assumes that all plies at 45° or 90° are considered to contribute 10 per cent of the strength and stiffness of an axial ply to the overall performance of the laminate in the "main" direction. For example, for a lay-up consisting of:

- 37.5 vol per cent fibres in the 0° direction,
- 50 per cent of ± 45° fibres, and
- 12.5 per cent of 90° fibres,

the 10 per cent rule sum gives the laminate strength and stiffness in the 0° direction as

(0.375 x 1 + 0.5 x 0.1 + 0.125 x 0.1) = 0.438

times the value of the relevant property measured on the unidirectional laminate.

Table 5.6: *Prediction of laminate efficiencies by the 10 per cent rule (Hart-Smith 1993)*

Laminate lay-up	Hart-Smith 10 per cent rule
UD	1
(0,90)s (Cross-ply)	0.550
(±45)s	0.100
[(±45, O$_2$)$_2$]s	0.550
[(±45$_2$, O$_3$,90)]s	0.438
(±45,90)s (quasi-isotropic)	0.325

Compression

The compressive strength of a composite element is influenced by the strength of the resin to a far greater extent than the tensile strength. The resin must hold the fibres straight to prevent them from undergoing local buckling or kinking under compression, and it also acts to help prevent failure through longitudinal splitting. At low stresses the stiffnesses in tension and compression are similar for CFRP and GRP, although the compression stiffness of aramid composites is often much lower than the tensile stiffness, (Piggott and Harris, 1980). Once damage starts to occur to the materials, FRP composites are weaker in compression than in tension. Designers should also be aware that when composite materials are subjected to damage by low-velocity impacts, the localised nature of this damage can often be a cause of premature failure under subsequent compression loading. In aircraft design a material characteristic referred to as compression-after-impact is often specified as an indicator of the ability of a composite to resist the effects of impacts.

The following table shows experimental tension and compression strengths for a variety of composite materials, and demonstrates clearly the difference between the tensile and compressive strengths.

Table 5.7: *Experimental tension and compression strengths for various laboratory made composite materials (Harris 1999, Piggott and Harris 1980)*

Material	Lay-up	V_f	Tensile strength σ_t GPa	Compression strength σ_c GPa	Ratio σ_c / σ_t
GRP	ud	0.6	1.3	1.1	0.85
CFRP	ud	0.6	2.0	1.1	0.55
KFRP (Kevlar)	ud	0.6	1.0	0.4	0.40
HTA/913 (CFRP)	[(±45, O$_2$)$_2$]s	0.65	1.27	0.97	0.77
T800/924 (CFRP)	[(±45, O$_2$)$_2$]s	0.65	1.42	0.90	0.63
T800/5245 (CFRP)	[(±45, O$_2$)$_2$]s	0.65	1.67	0.88	0.53

In designing for compression one should also consider the global behaviour of the element, and look at the possibility of buckling. Unfortunately the range of material combinations available means that there are no simple tables comparing slenderness and allowable compressive stress, so it is difficult to quantify this aspect of the design accurately.

The flexibility of FRP composite manufacturing allows the designer to maximise the resistance of sections against buckling, and more specifically against local element buckling. Troughing, edge flanging and double curvature can all be used locally to inhibit buckling.

Shear

There are two aspects of shear behaviour that require careful attention. First, when subject to in-plane stresses, the stiffness and strength of a laminate will reflect not only the load-bearing abilities of the fibre and matrix, but also the extent to which stresses are transferred into the fibres through the matrix and their interface. An illustration of the variation with fibre orientation of the Young's modulus (as given by equation 3) and the corresponding variation in the in-plane shear stiffness is given for typical unidirectional GRP and CFRP composites in Figure 5.2. A laminate of a balanced 0/90 lay-up possesses its greatest stiffnesses along the two fibre directions, but its in-plane shear stiffness is greatest at 45° to the fibres.

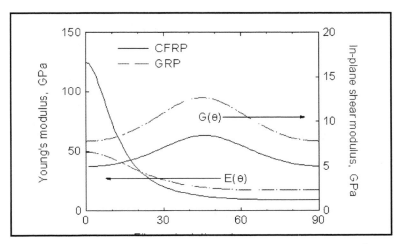

Figure 5.2: *Orientation dependence of the Young and shear moduli, E (θ) and G (θ), of unidirectional GRP and CFRP composites (Bryan Harris)*

Second, most laminated fibre composites contain planes of weakness between the laminations and along the interfaces, and in shear resulting from bending deformations, these planes of weakness can result in premature interlaminar failures and delaminations (Pagano, 1989). A property known as the interlaminar shear strength (ILSS) is used to provide an indication of this behaviour. The ILSS of high-performance laminates made from non-woven prepregs may be quite low, but the planes of weakness are less marked in composites manufactured from woven cloth or those containing short fibres, some of which lie at small angles to the plane of the sheet. Modern high-performance laminates are also often improved by through-thickness stitching with "z-direction" fibres. Since the ability of a composite laminate to support major structural loads depends on the efficiency with which such loads can be diffused into the fibres (and away from any local stress concentrators, like joints) by shear, it is important to ensure that the local shear resistance of the material is sufficiently high for the purpose.

Flexure

Flexure is really a combination of tension, compression and shear, and involves a complex inter-relationship between the fibres and the resin. At the most simple level it is acceptable to consider the flexural element in terms of maximum tensile and compressive stress in the outer fibres. Laboratory tests on composite beams in flexure have rarely shown the weakness in direct compression mentioned above and the apparent tensile strength is normally greater than its uniaxial tensile strength. By contrast a manufacturer has found that in tests on mast, in all cases but one, the failure was on the compression side first and the flexural strength was less than the axial tensile strength.

Note that flexural strength is seldom the limiting criterion, as stiffness more often dominates design.

Multi-axial loading

For most practical purposes, designers require models of behaviour that can predict failure under realistic combinations of stresses, rather than for the idealised uniaxial stress conditions under which most laboratory tests are carried out. There has been a great deal of research on complex-stress failure criteria. At least 20 models have been proposed, although the differences between many of them are quite small (Owen and Found, 1972), and designers remain in dispute as to the "best" model. The main difficulty in accepting one or other of the common design methods is that their validity can usually only be tested over limited ranges of combined stress. This is because of the complexity and cost of the test samples and test procedures for such experiments – for example, tubes under combined tension, torsion and internal pressure. Most of the existing failure criteria are in fact restricted to conditions of plane stress (thin plates) and some are only applicable to orthotropic materials. An exhaustive analysis was recently published under the editorship of Hinton, Soden and Kaddour (1998).

Fatigue

In common with all structural materials, the behaviour of composites under cyclic or repetitive loading must be considered (Owen, 1970, 1974; Talreja, 1987; Reifsnider, 1991; Harris, 1994, 1996). The fatigue behaviour of composites differs from that of, say, steel. With steel, failure tends to result from the intermittent propagation of a single crack, and the material even quite close to the crack is virtually unchanged. The inhomogeneous, an isotropic nature of composites results in fatigue damage in a general, rather than localised manner, and failure does not always occur through the propagation of a single crack.

Damage modes for composites include fibre breakage, matrix cracking, debonding, delamination and transverse ply cracking, and the predominance of one mode, or the interaction of several is dependent on the properties of the fibre and the resin. It is not certain whether composites exhibit the familiar "fatigue limit" known to users of steels. Few experimental results have shown any clear indication of such a limit, although in modern aircraft design the concept of a zero-growth threshold for pre-existing damage is now being used.

It is not safe to assume that fatigue can be ignored if working stresses are kept low (perhaps to avoid creep deformations). Stresses in the direction normal to the fibres could possibly be high as a result of load/fibre misalignment, which could lead to fatigue damage.

In order to minimise the risk from fatigue problems it is appropriate to make use of design data where available and sensible use of generous safety factors.

STRESS AND DAMAGE

Composites differ from conventional materials in the way in which they respond to stress. Under load, many local microstructural damage events occur – including resin cracks, fibre breakages, local fibre/resin de-bonds, and de-laminations. Depending on the particular laminate lay-up, the damage processes may begin to occur even at very low strain levels and the accumulation of damage continues as long as loading is sustained. This damage is usually widely distributed throughout the stressed composite, and in the early stages of life it does not seriously impair the load-bearing ability of the material. Moreover, this gradual accumulation of damage gives rise to changes in the material's physical properties that can be detected by means of suitable optical, dielectric, or ultrasonic techniques, which can therefore provide the opportunity for constant health monitoring of critical structures (as discussed in Section 5.2.2).

This complex mode of damage accumulation has important implications for the toughness of composites and for their fatigue response. Because of the interaction of the effects of these microstructural damage mechanisms and their energy-absorbing ability, many fibre composites exhibit excellent toughness and resistance to impact by comparison with conventional engineering materials. The structural use of reinforced plastics depends primarily on their mechanical response to load and also on their ability to absorb energy, either as a means of inhibiting crack growth or as a means of controlling the effects of impact. On the other hand, low-velocity impacts may cause local sub-surface delaminations, which can result in a reduction in the compression strength. Indeed, the compression strength after impact (CAI) is used as an indicator of the severity of impact damage sustained by a composite.

As damage accumulates in a composite under load, a critical point will be reached at which the material can no longer sustain the applied load and failure occurs. If the material is subjected to a cyclic stress, the residual strength of the damaged material will become equal to the maximum level of cyclic stress and a fatigue failure ensues. This process of "wear-out" of the material, which may be gradual throughout the life or may occur more rapidly at the end of the life, results in fatigue stress/life curves similar to those exhibited by metallic materials and design procedures are available for coping with this phenomenon. Composites with high elastic stiffness, which operate at low working strains (ie CFRP) have better fatigue resistance than less rigid materials like GRP and aramid composites.

In order to improve procedures for designing with this class of composite materials that accumulate damage under load, and permit their use beyond the point where damage is first initiated, the concept of damage-tolerant design has gained considerable attention in recent years (Dorey, 1982). The concept is not new: in effect it adopts the principle that the microstructural damage events, which can occur in the neighbourhood of a stress concentrator (a crack, for example) can have a beneficial effect in reducing the rate of crack growth in a manner somewhat analogous to that in which the plastic zone ahead of a crack tip in a metal can also inhibit crack growth.

Creep

Long-term deformation under stress, or creep, affects all materials to some extent, but some FRP composites must be checked for creep rupture. The exceptions to this are those FRP composites in which the creep deformation is almost completely prevented by aligned fibres of extremely high rigidity. In order to reduce the problem of creep the designer should generally reduce the permanent stresses to such low levels that the creep is reduced to negligible levels. This can be done by limiting the permanent service strain level to avoid creep rupture.

Toughness

It is easy to be misled by statements about toughness because of the lack of any universal standard in defining and measuring toughness across the engineering spectrum. The most highly developed concepts of fracture mechanics used routinely in aerospace and pressure-vessel engineering, for example, appear to be largely ignored by civil engineers. The formal fracture-mechanics concept of a "fracture toughness" (for example see Broek, 1986) are also difficult to apply to fibre composites, because their fracture modes do not easily fit the models used in formal fracture mechanics.

A useful property with which to compare materials, is the work of fracture – the energy required to cause a sample, usually containing a sharp notch, to break into two pieces, expressed in relation to the sample cross sectional area. As such, it is related to the familiar impact energy properties known by the names of Charpy and Izod. A few comparisons can illustrate certain advantages of composites. Tough steels have works of fracture of the order of 100 kJm^{-2}, while brittle materials like ceramics and glass have works of fracture of only a few Jm^{-2}. Resins like epoxies and polyesters exhibit values of a few hundred Jm^{-2}. However, when fibres (even brittle fibres like glass and carbon) are combined with resins in composite materials, they usually have works of fracture of between 10 kJm^{-2} to 100 kJm^{-2}, depending on lay-up and composition.

This unusual feature is a result of the fact that many composites fail in very complicated ways by the occurrence of a multitude of micro-scale failure events (fibre breakage, resin cracking, fibre/matrix debonding, delamination, etc), which boost the toughness, defined as above, in the same way that plastic deformation ahead of a crack tip in a metal increases its toughness. This is a major advantage of fibre composites. Typical values of the impact energies of some composites are given in Table 5.6.

Thermal expansion

The coefficients of thermal expansion (CTEs) of composites are anisotropic, like their elastic properties, and there are several important consequences for designers (Holliday and Robinson, 1977; Halpin, 1992; Eckold, 1994; Hancox and Meyer, 1994). The CTEs parallel with and transverse to the fibres in a unidirectional lamina are given by the expressions (Schapery, 1968):

$$\alpha_1 = \frac{E_f \alpha_f V_f + E_m \alpha_m V_m}{E_c} \tag{6}$$

$$\alpha_2 = (1 + v_m)\alpha_m V_m + (1 + v_f)\alpha_f V_f - \alpha_1(v_f V_f + v_m V_m) \tag{7}$$

where α_1 and α_2 are, respectively, the axial and transverse CTEs. In polymer-matrix composites α_1 is strongly fibre-dominated and falls much more rapidly than a mixtures-rule prediction as V_f increases, whereas α_2 initially rises slightly before following an approximate mixtures rule between the fibre and matrix values. The maximum difference between the CTEs occurs at a volume fraction of about 0.2 for a typical E-glass/epoxy laminate. A consequence of the expansion anisotropy is that laminates containing plies at different angles will distort in unexpected ways unless care is taken to ensure that the constraints are always matched in designing the lay-up of the material. A single pair of unidirectional plies of an E-glass/epoxy laminate of V_f = 0.6 bonded together at 0° and 90°, for example, would behave like a bimetallic strip during unconstrained cooling from the curing temperature because their expansion coefficients, given by the above two equations, would be 7.9 x 10^{-6}K^{-1} along the fibres and 5.6 x 10^{-5}K^{-1} normal to the fibres. During the constrained cooling of a finished component from its manufacturing (or curing) temperature residual stresses will be set up inside the material. Continuing with the glass/epoxy example, a balanced (0,90$_2$,0)S laminate would not distort on cooling, but there will be a residual thermal strain of nearly 0.5 per cent transverse to the fibres. In a composite with a tough matrix resin, this would not cause problems, but such a level of strain could well be high enough to cause cracking in the resin or at the fibre/matrix interface in a composite with a brittle (low extensibility) matrix. It is important to note that, on account of their molecular structures, carbon and aramid fibres exhibit negative expansion coefficients along the fibre axis, as indicated in the list of properties in Table 5.8.

Table 5.8: *Some typical thermal expansion properties for polymer composites (Hancox and Meyer, 1994)*

Material	Vf	CTE. 10^{-6}K^{-1} (parallel to fibres in UD, or any direction in isotropic lay-up)	CTE, 10^{-6}K^{-1} (transverse to fibres in UD)
E-glass/epoxy	0.60	7.9	56
E-glass polyester SMC	0.18	2.0	
E-glass polyester DMC	0.12	5	
E-glass CSM/phenolic	0.22	15	
T300 carbon/epoxy	0.60	0.3	37
P75 carbon/epoxy	0.65	-1.1	32
Kevlar-49/epoxy	0.60	2.1	69

Apart from the problems relating to expansion anisotropy, residual stresses and cracking described above, designers would also need to exercise care in designing structures combining FRP composites and other more conventional materials in order to avoid incompatibilities resulting from thermal expansion mismatch.

Safety factors in design

Current codes in Europe require that structures are designed separately for the serviceability limit states and the ultimate limit states (failure). Serviceability limit states include deflections and vibrations. High performance composites made with carbon or kevlar are expensive so they will only be used where advantage can be gained from their low weight and high stiffness. This means that they will, almost exclusively, be designed primarily for serviceability requirements where the load and material factors (see below) are 1. Because they fail without yielding, engineers take a conservative view of the factors of safety against failure.

For the ultimate limit state factors of safety in structures are included as load factors γ_f and material factors γ_m. For composites we would recommend following the principles used in the aluminium code BS 8118 since, like aluminium, FRP composite are used for a greater range of products than just building structures and there is a greater range of material properties. In this code the factors for loads and materials are clearly separated and are different to the factors in the steel code BS 5950 where some of the material factors are included with the load factors. The γ_f factors can be used as defined in BS 8118 but the material factors need further consideration.

Material factors have been put forward in the *Eurocomp Design Code* (Clarke 1996). The overall material factor is the product of a fibre type factor and a manufacturing process factor. $\gamma_m = \gamma_{mf} \times \gamma_{mm}$. While we would agree with this approach for establishing an overall factor of safety we do not believe that there is sufficient experience with the materials to lay down definitive numbers. One has to think of the implications of failure and each situation should be evaluated on its merits. We are suggesting a framework in which this can be done.

The aim of the use of material factors is to arrive at a safe design stress. The starting point is the estimation of the characteristic failure stresses for the different loading modes and directions. This can be done in three ways:

1 By calculations based on the amount of fibres and their orientation in the composite. This can be done using the Krenchel formula and the Hart-Smith rule described above, or by a more complex calculation. This probably does not deal with compressive strength very well.

2 By testing coupons cut from samples that have been made in a similar way to that intended for the object to be made. The tests would include compression and shear strengths.

3 By testing the actual object or a large piece that was geometrically similar.

The characteristic failure stress from tests would normally be assessed as the mean of a reasonably large number of tests, minus 2.33 standard deviations. This would give a value below which 1 per cent of results would fall. If method 1 is used a factor, γ_{mm}, related to the method of manufacture and its impact on the strength of the component should be applied. This factor needs to be established by tests on actual components in a similar way to methods 2 and 3. The result should be to arrive at comparable nominal failure strengths to those predicted by using methods 2 and 3 directly.

Table 5.9 *Suggested values for γmm*

Method of manufacture	Partial factor on calculated strengths γ_{mm}
Machine controlled system eg pultrusion	1.1
Vacuum infusion, RTM	1.2
Wet lay-up with good quality control and vacuum consolidation	1.3

The other material factors are intended to allow for the fact that the failure strength of a manufactured product may be less than that predicted by factored calculations or material tests. The safe design stress adopted will also be affected by the mode of failure of the particular fibres and construction of the composite product (γ_{mf}). The amount of damage the product may experience over its working life and the consequences of failure will also have to be taken into account by a "damage" factor (γ_{md})

With steel structures the failure point is considered to be the yield point. The material has considerable ductility beyond this and the actual failure stress is some 40 per cent greater than this. By contrast most FRP composites have little, and often poor, ductility that tends to vary inversely with the strength. GRP in the form of chopped strand mat has randomly orientated fibres and hence has low strength but can undergo considerable deformation giving visible and sometimes audible evidence of the development of damage. GRP laminates made from woven cloth in 0-90 and 45-45 directions will have greater strength and much less ductility while unidirectional pultrusions have even more strength and no ductility. Kevlar or especially carbon fibres have higher stiffness and less elongation at break so there is even less warning of failure which can be explosive if the stress is high enough.

Because of this the higher stiffness fibres and constructions should have a higher material factor.

Table 5.10 *Suggested partial material factors for directional lay-up composites in tension*

Material	Partial factor for fibre type and mode of failure γmf
Carbon Fibres	> 2.0
Ultra-high modulus carbon fibre	>2.5
Aramid fibres	2.0
Glass fibres	2.0
Glass fibres, CSM with good quality control	1.5

Note: Compression characteristic strengths and their associated partial safety factors may well be different from those above.

The damage factor γmd takes into account the likelihood of the failure strength being reduced over its working life. The strength of glass fibres can be reduced by contact with water. Stress fluctuations can cause delamination, debonding etc, which will reduce the strength over time. High stiffness fibres are less susceptible to this effect. The low ductility composites will be susceptible to stress concentrations at drilled holes, lay-up flaws, discontinuities in section at glued joints etc. A suggested damage factor for glass fibres when wet is in the range 1.2 to 1.4.

Experienced manufacturers of structural products, such as carbon fibre masts or wind turbine blades, rely on a library of material tests based on the actual lay-up and fibre types that they regularly use. From these they apply partial safety factors to derive safe design stresses for the different directions of loading and applications. We would advocate following this approach and discussing the design stress levels with the fabricator.

Remember that particularly high performance materials are mostly used for products where strength is not the most important consideration. Most FRP composite products are designed for serviceability, which generally means stiffness. Mass produced products can be prototype tested and strengthened locally if necessary. Other applications such as the strengthening of reinforced concrete beams have been type tested so that appropriate safety factors are established. Kits for this application are now available off the shelf and come with guidance for their use. The same should happen for pultrusions if they become commonly used as structural elements.

It is the one off special structural elements where the safety against failure becomes an important issue, for example a mast to go on a high building. In this Carbon or Kevlar fibres would only be used where low weight and high stiffness are required and hence strength may not be the governing factor. As with other structural applications the implications of sudden fracture should be taken into account and additional security measures added if possible to avoid for example the broken part falling into a place where there may be people.

As an example, in work for London Underground, a jack arch was strengthened with a CFRP beam, and this was designed with a safety factor of 2.5, recognising the high cost of failure (Moy, 1999).

Further, in interim guidance (to be published) from the Institution of Civil Engineers, a number of safety factors are given for reinforcing systems using FRP composites. These cover the range 1.5 to 3.6 for the main reinforcement, and 1.1 to 1.8 for the material modulus on ultimate limit state loads. The Concrete Society has also been working on guidance in this area, (Concrete Society, 2000).

Summary of FRP composite properties

When considering fibre orientation in practical composites, materials can be broken down into three major hierarchical classes as follows, and as shown in Table 5.11 opposite.

1   Composites with unidirectional fibres are highly anisotropic, with very high strength and stiffness in the direction of the reinforcement. Away from this direction, properties fall away rapidly until at the orientation perpendicular to the fibres they become similar to those of the resin matrix. Composites reinforced with unidirectional fibres are usually only used in situations where the loads will only be applied along the line of the element. Typical structural uses include pre-stressing tendons, shear stiffeners, plates for beam strengthening and I beams.

2   Composites containing laminates or woven materials have fibres at different orientations and, while still anisotropic, have at least some strength in every direction. A great advantage of composites is that when building up a composite using layers of fabric, the orientations of the fibres can be specifically set to resist the applied loads. Composites reinforced with laminates have very versatile structural performance and can be designed to resist a range of loads. Folded-plate structures and retrofit strengthening plates are examples of situations where laminated reinforced composites may be used.

3   Randomly oriented fibres produced by spraying or using chopped strand mat are used less often for specifically structural purposes, and are more likely to be used for non-structural elements such as cladding panels. While such elements behave in an isotropic manner, their strength and stiffness are significantly lower than that of either of the previous two categories.

Table 5.11: *Hierarchical characterisation of types of composite used in construction*

Performance or quality level	Type of composite	Approx. fibre content (V₁)	Nature of mechanical properties	Typical manufacturing procedures	Typical applications in construction
LEVEL 1	Continuous fibres, unidirectionally aligned in the composite	0.6–0.7	Highest performance in a single axis: highly anisotropic	Pultrusion Prepreg and vacuum bag	Pre-stressing tendons, concrete rebars, I beams and other sections, shear stiffeners, racking, retrofit-strengthening
LEVEL 2	Continuous fibres, arranged in several directions to meet specific design requirements	0.40–0.65	Medium high performance; anisotropy controlled according to requirements, eg good properties in two directions	Filament winding, hot pressing between shaped dies, RTM, prepreg or hand lay-up	Pipes and vessels with a symmetry axis, plates and shaped components; hollow tanks and vessels, folded-plate structures, cladding panels, simple or complex mouldings
LEVEL 3	Short or chopped fibres, randomly arranged in two or three directions. Fibres may be blended in with the polymer before moulding, or may be in the form of a mat (chopped-strand mat)	0.01–0.40	lowest performance; material isotropic or approximately so in two directions, possible anisotropy in third direction	hand and spray lay-up, RTM compression moulding, injection moulding	simple or complex mouldings, domestic services ducting and rainwater goods, concrete form-work

The Hart-Smith rule deals with the first two classes, but does not address the third, random arrangement of fibres. Krenchel's model (1964) suggests that for randomly arranged fibres a factor of 0.375 can be used in equation 5 for mouldings with planar-random fibre distributions and 0.167 for thicker mouldings with a 3-dimensional fibre array. Again these factors should be multiplied by the value of the relevant property measured on the unidirectional laminate. A schematic illustration of the relative tensile properties of classes of reinforced polymers within this hierarchical scheme is shown in Figure 5.3.

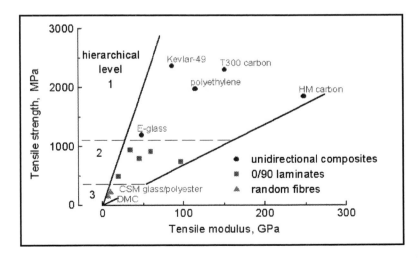

Figure 5.3: *Strength and elastic modulus of some typical reinforced plastics. The hierarchical levels are those referred to in Table 5.11. The grouping for the level 2 materials is the same as that indicated for level 1 (Bryan Harris)*

There are numerous other factors affecting the strength of a composite section, although it is neither possible nor necessary to go into these in detail in this publication. Another factor that has a significant effect, is the length of the embedded fibres. Fibres have a critical length below which a tensile force applied to the fibre will break the bond between the fibre and the resin rather than break the fibre itself. For a composite with fibres averaging 10 times this critical length the strength of the section is approximately 95 per cent of a section reinforced by the equivalent continuous fibre. Unfortunately the critical length is a function of the strengths of the fibre and the bond between the fibre and the resin, so cannot easily be listed. Table 5.12 summarises key data for a number of typical structural FRP composites.

As given elsewhere in this report, these values are either ranges or typical average properties. The table is incomplete because not all of the chosen properties have been reported for all of the materials.

Table 5.12: *A comparison of the principle mechanical properties of a range of common FRP composite materials*

| Material | Fibre content V_f | Density 10^3 kg. m^{-3} | Elastic moduli (GPa) | | | Strengths (MPa) | | | Interlamiar shear strength (MPa) | Impact energy[2] kJm^{-2} |
			Young's	flexural	shear	tension	flexure	compression		
Glass-fibre reinforced polymers										
a) thermoset matrices										
pultruded rod	0.70	2.09	54	(54)		1600		500–700		>300
ud laminates (prepreg)	0.55–0.65	1.8–2.0	35–50	30–40	4–6	600–1200	900–1400	500–700	70–90	220
0/90 laminates (prepreg)	0.5–0.6	1.7–2.0	20–35			300–500			72	125
0/90 laminates (woven)	0.30–0.55	1.5–1.9	10–25	18–22	5	200–360	500–600	240–320	30–60	30
CSM/polyester/continuous sheeting	0.12–0.25	1.3–1.9	5–15	12	3	50–200	280	150–220	12	35–60[3]
SMC	0.12–0.25	1.3–1.9	12–14	10–13		50–100	155–255			20
DMC	0.05–0.15	1.2–1.8	5–13	7–14		20–40	40–100			18–24
b) thermoplastic matrices										
ud E-glass/PPS	0.57		52	42		835	1150	800	43	
0/90 woven E-glass/PPS	0.60		24	21		248	366	297		
RTP (glass-filled Nylon)	0.20	1.1	5–11	5.5–6		160–190				
GMT (glass-filled polypropylene)	0.13	1.13	5.5	4.5		85	140			
Carbon-reinforced thermosets [1]										
ud laminates (prepreg)	0.57–0.6	1.6–1.8	130–300		5–8	600–1200			60–90	50–100
0/90 VARTM laminate (non-woven)	0.60–0.65		72	72		900	1149		70	
0/90 laminate (woven)	0.50–0.60	1.53	60–70	60–70	5	600–900	800–1100	500–700	57–65	
Aramid-reinforced thermosets										
ud laminates (prepreg)	0.60	1.38	76		2	1200–1400		280	60–83	
0/90 laminate (woven)	0.50	1.33	31		2	517		172	70	

[1] properties depend strongly on fibre type   [2] measured on notched samples
[3] depending on thickness

Data from various sources, including Eckold (1994), Hancox & Meyer (1994), Harris (1999), Vetrotex technical literature, Delaware Composites Encyclopedia (1990)

Detailed analysis and lay-up design

This section of the report has given an overview of the principle factors affecting the structural behaviour of composites, however it has not given sufficient information to carry out a detailed design and analysis of such a section. In order to take the design to a more detailed stage the designer could obtain a specially written finite element computer program that calculates the strength of sections with particular lay-up and fibres. Designers could also consult other literature on the subject.

The aim of the previous sections is to enable structural engineers to carry out scheme designs of FRP components. This coupled with an appropriate specification should be sufficient to obtain tenders and select a fabricator. Unless the engineer is familiar with the detailed process of selecting the fibres and designing the lay-up of the composite, we suggest that as part of the contract the fabricator is required to undertake this work. Specialist FRP fabricators will often have an in house design capacity and can design the lay-up of components in a manner that both takes care of the structural aspects and ensures that the component can be manufactured with the maximum efficiency. If they do not there are a few consultants who specialise in FRP composite design who can help the fabricator with this process.

Connections/joining

Considerable work on the joining of FRP composites has been carried out in another CIRIA project, and published as Project Report 46 (Hutchinson, 1997). It covers the design of joints in great detail, so this report only gives an introduction. Coverage of joining and bonding is also given in the publications of Adams and Wabe (1994), Kedward (1981) and Matthews (1983, 1987).

The virtue of composite materials is that they offer great scope for revolutionary, as opposed to evolutionary design. Multi-component structures can often be redesigned to take advantage of the capability of several of the common manufacturing processes to fabricate a single monolithic end product where the material and the structure are produced in one operation. However, while it is rarely a good idea simply to carry out piece-meal substitution of composites parts for existing parts made of conventional materials, there are occasions when it will be necessary to combine standard composites products, such as plates and pultrusions, in order to build up more complex structures. This may be necessary to give scope for disassembly, for transportation or repair, for example.

The Eurocomp Design Code (Clarke, 1996) identifies three classes of connections involving composites:

1 Primary structural joints that carry major strength and stiffness to an assembly for the whole-life of the structure (eg stressed-skin panels, beam/column connections, concrete to FRP composite joints).

2 Secondary structural joints, failure of which would cause only local failure without compromising the entire structure (eg cladding panels and modules in building structures).

3 Non-structural connections, which include movement joints that could incorporate a gasket or flexible sealants of various kinds, the main purpose of which is to exclude the external environment.

Both mechanical and bonded joints are used for FRP composites, depending on requirements, and hybrid joints combining the two are often used. Furthermore, there have been recent developments in "snap" joints for rapid assembly – touched on briefly in Section 4.3.3. These involve pieces prepared in such a way as to allow secure construction to be carried out in a few seconds, and subsequent disassembly later.

MECHANICAL FIXINGS

Most types of conventional mechanical fasteners are used – screws, rivets and bolts – and in appropriate cases screwed inserts can be moulded into one of the joint components. For simple mechanical joints, the advantages are that no surface preparation is needed, disassembly can be achieved without damage, and there are no unusual inspection problems. The disadvantages are that fastener holes cause stress concentrations and the fasteners themselves may involve a significant weight penalty, limitations that do not apply for bonded joints.

The basic approach to the use of mechanical fasteners is the same as for metallic joints, except that the anisotropy of the composite must be allowed for. To develop the maximum possible strength, through-thickness restraint must be provided by the fastener, so that bolted joints are stronger than riveted ones, which in turn are better than screwed joints. The dimensions of the joint must be carefully chosen to ensure that bearing failure occurs as opposed to the weaker tensile or shear modes, and these dimensions are determined by the properties of the materials being joined.

ADHESIVE FIXING

Adhesively bonded joints, on the other hand, avoid some of the problems of stress concentration, but have their own disadvantages. Careful surface preparation is required. Joint integrity is sometimes difficult to confirm by inspection methods, joints can be severely weakened by hostile environments and disassembly is not possible without component damage.

In adhesively bonded joints, the joint is strengthened by increasing the stiffness of the surface to be fixed, and reducing the elastic modulus of the adhesive itself, most frequently an epoxy resin. Joints must be designed so that the peel (bending) stresses are kept to a minimum. For the transfer of high loads, scarf or stepped joints must be used, although these are more difficult to make than the basic single- and double-lap joints. A ductile adhesive is preferred to a brittle one because both the static and fatigue strengths are improved in this way. However, since a bonded joint relies on stress transfer though an unreinforced adhesive resin, creep is always likely to be a problem.

It can also be beneficial to use a hybrid joint system, where both an adhesive and a mechanical joint are used. This utilises the strengths of both types of joint. For example, the adhesive joint benefits from a bolted fixing in keeping it secure during the curing process, and in resisting peeling and shear forces. The adhesive helps to spread the load, reducing the impact of stress concentrations around the mechanical fixing. A development of this principle is shown in the recently patented idea of the AdhFASTTM joining system developed at The Welding Institute, which uses a mechanical fastening device to permit dry assembly, simultaneously maintaining an appropriate gap between the components into which a two-part resin adhesive can then be injected through the fastener itself. This system provides excellent control of the bond-line thickness, the spread of the adhesive and, therefore, of the quality of the joint, bringing the added advantages of improved performance and cost savings.

COMPATIBILITY
Some consideration must also be given to the joining of FRP composites to other types of material in the structure. The bonding of FRP composites to other engineering materials, such as concrete, steel or aluminium, can be satisfactorily achieved, provided proper control of the choice of adhesive and the preparation of all surfaces is exercised. Epoxy resins are already widely used for bonding concrete, aluminium alloys and (to a lesser extent) steels, to themselves and there is no problem bonding the same materials to the epoxy or polyester resin surface of a FRP composite, therefore.

However, problems may arise when bonding together materials that possess quite different physical properties. In particular, stress concentrations may occur at joints between two materials, which possess markedly different elastic moduli and thermal expansion characteristics, and the joints must therefore be designed so as to minimise these stress concentrations, or at least to ensure that they cannot act to destroy the joint itself.

Chemical incompatibility may also sometimes be a problem. Two real examples will illustrate this. If a polyester-based FRP laminate containing residual free styrene is brought into contact with another polymer, such as polystyrene in bulk or foam form, the latter may be softened by the styrene in the FRP composite and lose its structural, optical or insulating properties. And if a CFRP laminate (reinforced, it will be remembered, with filaments that are electrically conductive) is attached to a steel or aluminium structure in such a way that permits electrical contact between the fibres and the metal, a galvanic cell may be set up, which could result in severe corrosion of the metallic part of the joint. Again, attention to detail can avoid these problems at the design stage.

5.2.2   Durability

Weathering

Many clients expect their buildings to have a low maintenance life of 50 years or more, so durability is of particular importance for architectural applications. Weathering primarily affects the surface appearance. Colours may fade, the surface will lose its gloss and with some materials there can be excessive dirt retention. The best way to gain confidence with the performance of FRP composite products, is to look at examples that have been in use for a long period. A number of the case studies in Chapter 3 are included for that purpose. Generally, external GRP cladding units have performed well. Several of the examples are 25-years-old or more and are still looking good, although some have aged badly and this is discussed below. This should be compared with the competing materials, many of which are replaced completely on a shorter time-scale than this, or require regular re-painting.

As with paints and plastics UV light degrades the resins used in the manufacture of composites. This degradation is reduced by the use of stabilisers, fillers and pigments. Some pigments, particularly blue, tend to fade, so they have to be selected with care. The degraded resin leaves a white powder, which will leave light patches on a dark coloured surface. The discoloration effects can be reduced by cleaning. It can be avoided by using light colours that can tolerate the loss of shine and whitening. Light colours also keep the surface temperatures down, which reduces internal damage to the composite that can sometimes be caused by high thermal gradients. As in other external applications of polymers in general, titanium dioxide is most frequently added as a UV filter. Further discussion of weathering is given by Leyton (1999) and Scott (1970).

Repair

If a cladding panel gets damaged by, say, a vehicle reversing into it, it can be readily repaired with glass cloth and resin. The problem is to maintain the smooth line of the surface and to get a colour match. The surface can be smoothed by applying the repair below the level of the original surface and making it up with filler. The whole panel can be re-coated with gel-coat, which will ensure a consistent colour over the whole panel but may not match with the others.

Environmental degradation

The long-term response of a fibre composite to its environment is complex. High-performance materials like CFRP and aramid-fibre composites are likely to remain stable under conditions relating to civil engineering structures (modest maximum operating temperatures) because modern versions of both types of fibre are relatively insensitive to moisture or temperature. By contrast, the ordinary glass fibres present in GRPs are attacked by moisture, acids and alkalis, and the filaments can easily lose their strength if they are not adequately protected by the surrounding resin. Detailed discussion of these issues is given in the publications of Aveston and Kelly (1979), Dorey (1982), Hogg and Hall (1983), Pritchard (1998) and Jones (1999).

Matrix resins have different degrees of resistance to moisture and temperature and must be selected according to requirements, in order to control environmental deterioration as far as possible. Moisture diffuses through most resins, and if it reaches pores or fibre/matrix interfaces or interfaces between the plies of a laminated composite, or if it penetrates pre-existing cracks in the resin, reductions in mechanical properties of the composite will occur. If liquid is able to penetrate near-surface cracks or bubbles or regions of the resin that contain contaminants, blisters may occur, which not only ruin the appearance of the material but also increase the rate of access of the moisture to the interior. A detailed discussion of these and other defects is given in the *Cystic Polymer Handbook,* Scott Bader (1994).

In order to maximise the durability of composite structures for use in corrosive environments, complex layered structures are built up, which include resistant surface coatings and diffusion-barrier layers in addition to the main load-bearing sections. Boats and sewer pipes are good examples of applications where specialised systems have been developed by manufacturers, but the surfaces of such systems, far from being "maintenance-free" as is commonly stated, must be properly maintained.

The effects of moisture and elevated temperature on matrix resins are similar in the sense that both lead to reductions in the strength and stiffness of the resin. If the behaviour of the composite system is matrix-dominated rather than fibre-dominated, this can lead to increased creep rates over time. All resins creep to some extent, even at ambient temperature. Creep also affects the stability of the glass fibres in a time-dependent manner through a process sometimes referred to as static fatigue, or environmental stress-corrosion cracking. The basic equations governing both the static fatigue response of glass fibres and the rate of moisture diffusion into resins can be combined to provide approximate life-prediction procedures.

Maintenance

It is a general strength of FRP composite materials that when well designed and manufactured they require limited maintenance. However many conditions will need periodic cleaning, and checking the condition of joints and sealants. Where there are structurally significant elements, checks for evidence of any cracking will be required.

Most composite materials sustain damage when loaded, either during proof testing or in service. Due to the nature of the composite material, a structural composite component is usually able to tolerate a considerable amount of damage accumulation without losing its load-bearing ability. One consequence of such damage is that it may allow ingress of an external aggressive agent (eg water) into the material, accelerating the rate of deterioration of the interfacial bonds and the load-bearing fibres. Manufacturers and users need practical techniques for the detection of these defects and damage. Much of the technology previously developed for metallic engineering materials and structures has been transferred across, with appropriate modifications, for use with fibre composites. It is not appropriate, in this report, to give a detailed description of available methods, but a list of the most common techniques is given below. Further information can be obtained from Harris and Phillips (1983), Summerscales (1987), Harris and Phillips (1990 a, b and c) and Harris (1994).

The most common NDT methods in use at present are:

1   Optical inspection, by reflected or transmitted light, often with the assistance of dye-penetrants.

2   X-ray methods, contact or projection radiography, with radio-opaque penetrants if cracks are being sought.

3   Microwave methods, eg by surface measurement of the dielectric constant.

4   Dynamic mechanical analysis, by observing the transient response of the structure to a mechanical impulse, eg "coin-tap" testing.

5   Thermal imaging, with the use of liquid crystal coatings or with infra-red imaging systems (pulsed or continuous).

6   Ultrasonic techniques, most commonly by through-thickness attenuation measurements (C-scanning). Conventional C-scanning is a slow process, but new techniques for real-time operation are now coming onto the market.

A new feature that is becoming possible now that electronic systems are so inexpensive is the self monitoring structure. Optical fibre sensors, either embedded or surface mounted, offer important possibilities for on-line application in the development of "smart" structures. In the future we expect to see components that are able to measure their own performance and report on this to a central system, allowing problems to be dealt with at an early stage. Of course all materials could allow this, but it is well suited to a moulded material like FRP composites.

Examples of failures

There have been a number of FRP composite systems that have failed in the past. It is important to learn from mistakes made, rather than to ignore them. The following examples are taken from work at the BRE (Halliwell, 1999), and involved re-visiting a number of 25-year-old GRP clad buildings. The lesson from these is not to avoid the materials, but to ensure that the design is carried out properly, and that the appropriate materials are built and installed correctly.

Market place roof, Leicester

The roof of the market place in Leicester was built in a similar style to the previous "traditional" covered market. It consisted of truncated reddish-brown GRP pyramids on a steel framework. It was designed for modern visual appearance, to be lightweight and resistant to the environment, with limited maintenance.

The roof was taken down in 1991. The major problem was leakage from the four corner joints in the pyramids. It also discoloured badly and was a very dull light grey/slate colour when removed, but there was no evidence of embrittlement.

The opinion of the architects was that the units were under-designed and could not stand the thermal stress imposed on them, causing cracking at the angles which then let in water. There were also several cracks in the gutter areas. Structurally the roof was sound and was in no danger of falling down.

Figure 5.4: *Leicester Market place new roof in 1970 (Leicester Markets)*

Figure 5.5: *Roof prior to demolition c1990 (Leicester Markets)*

Morpeth School, London

The roof for an extension to parts of the school was built with GRP, both solid areas and "translucent" skylights. The expectation was of low-maintenance, and the materials were chosen for their appearance.

After 25 years several problems were apparent. The translucent material had darkened significantly, and the material was found to be significantly more brittle. In some areas the surface had degraded badly (see Figure 5.6) and when walking on the roof, cracking noises could be heard, although there was little visual evidence of cracks. Because of these the roof areas were taken down in 1997.

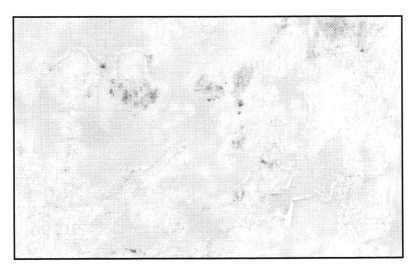

Figure 5.6: *Morpeth School London, built 1973 – detail of surface degradation (BRE)*

Figure 5.7: *Morpeth School (BRE)*

Services station building, Newport Pagnell

The walls of the restaurant at this service station were made from profiled single skin GRP, with a buff coloured gel-coat and resin, to meet the desired performance and appearance.

Early inspection showed that there had been problems in manufacture, with the gel-coat not covering the surface at corners and other exposed areas. The effects of this have become more pronounced over time.

In 1997 the panels were dirty and had lichen growth on them, and some uneven fading of the colour. Many panels had some crazing of the surface, and the cracked areas with no gel coat had the worst lichen growth.

Overall the appearance of the building has not weathered well, although it is still fully functional.

Some areas had been painted recently, and were in good condition. Others had been painted earlier, but the paint had peeled. Little repair work had been carried out, but some areas appeared to have been cleaned with an abrasive, leading to a scratched surface and more rapid dirt build up.

Figure 5.8: *Newport Pagnall Services building 1973 (BRE)*

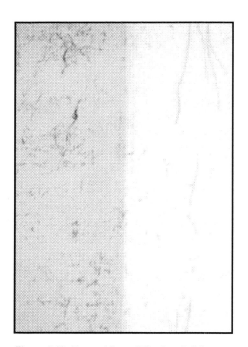

Figure 5.10: *Newport Pagnall Services building, showing surface degradation*

Figure 5.9: *Newport Pagnell Services building showing damage to window sill (BRE)*

5.2.3　Fire

Specially formulated composites are employed widely in the offshore industry for fire protection applications. They are used to protect pipe and ductwork and can form a major percentage of the total construction materials in new unmanned platforms. This means that designing FRP composites for fire can be done in one of the most demanding engineering conditions, from a fire perspective, and so it must also be possible in other situations.

This section introduces the subject of designing FRP composites from a fire perspective. It starts by outlining the different approaches that are possible and explains briefly the tests that are carried out, and what these can tell us. Finally it addresses the different materials (particularly the matrices) and how they perform, with and without the addition of fire retardants.

Other sources of information include: Berry (1990), Clark J L (1996), Concrete Society (2000), Burling (1999), Dodds *et al* (2000).

The approach to design

REGULATION-BASED
Regulations in most countries provide "deemed to satisfy" requirements for the fire performance of materials to ensure that they do not promote the spread of fire or contribute significantly to it. In order to demonstrate compliance with the requirements of documents such as Approved Document B (The Building Regulations 1991), materials to be used are required to have been tested in accordance with the relevant part of BS476 "Fire Tests on Building Materials and Structures". The approach of following the prescription is the simplest intellectually as it describes exactly what to do, but it can involve excessive cost for the client on account of over-specification.

PERFORMANCE-BASED DESIGN AND FIRE ENGINEERING
Performance-based design often differs significantly from the deemed to satisfy regulation-based approach, in that the requirement is set down, but without the method to achieve it. Examples include the time that a wall will resist a fire, or the time and fire size for a structure to stay intact. In the design of a building it may be necessary to have fire resisting load bearing elements, but through a performance based approach, it may be possible to prove that a particular element of construction is structurally redundant in the fire case and therefore does not require fire protection.

The performance-based design considers the design of the construction as a whole and does not focus on materials in isolation. Real structures perform better than individual elements of structure when involved in a fire. It can therefore be determined that a material, which achieves little fire resistance in a standard test, may achieve an adequate level in a real fire when the cumulative effect of considering real fires and the performance of the structure as a whole are accounted for.

Fire engineering is the performance-based approach to the design of the fire safety strategy in a building. The approach is usually used to demonstrate an adequate level of fire safety when an element of the design does not meet the recommendations contained within prescriptive guidance. It can be used to incorporate materials into the building design that do not meet the prescriptive recommendations by showing that there is only a small risk from doing so.

For example, a building is being constructed and a feature wall/exhibit, which constitutes a large surface area is being incorporated into the design. In order to comply with the recommendations of Approved Document B (The Building Regulations 1991) the surface of the feature will be required to achieve a minimum of class 1 surface spread of flame when tested to BS 476 Part 7. This is to be one of the main features of the building and will be to a high quality finish and an interesting shape and therefore is to be constructed from GRP. Owing to the requirement for a high quality gloss finish and the limited time available to fabricate the feature, the deemed to satisfy requirements cannot be met. Therefore a fire engineered approach is used to demonstrate that if ignited, the feature does not constitute a significant risk to life safety.

A typical strategy for incorporating such a material into building would be to prove minimal ignition sources and through effective building management, to maintain a fire sterile area surrounding the feature.

Fire testing

Any new product will almost certainly need to be fire tested, so a range of basic properties should be available to the designer from the tests discussed in this section.

FIRE RESISTANCE

Fire-resistance testing in the United Kingdom is usually conducted in accordance with the specification of BS476 parts 20 to 23 tests. However, these are only a small selection of the fire tests available for determining the fire resistance provided by an element of construction. Fire resistance can be tested in furnaces, by pool fires and by burner tests.

Rapid changes are taking place in fire testing procedures throughout Europe. Many of the current standards, to which materials are required to be tested, are to be changed. The replacement for BS476 parts 20 to 23 and part 31.1 have already been released and it is recommended that all new materials to be tested for fire resistance should be tested to the new standards. The new European standards for fire resistance are listed in the table below.

Table 5.13: *Equivalance of different standards*

Current British Fire Test	New European Fire Test Equivalent
BS 476 Part 20	BSEN 1363
BS 476 Part 21	BSEN 1365
BS 476 Part 22	BSEN 1364

It is worth noting that the new test standards for fire resistance are very similar to the currently referenced BS476 test series.

FIRE REACTION

Fire reaction tests determine how materials will perform when introduced to various types of thermal assault. These tests measure material properties such as heat release rate, combustibility and surface spread of flame, which all affect the rate and severity of fire growth within a building. BS476 parts 3, 4, 6 and 7 are the relevant British fire tests for determining the reaction of a material in a fire scenario.

AD HOC *FIRE TESTING*

An *ad hoc* fire test is a test that attempts to replicate the proposed end use of a material. Rather than a generic fire test designed to give a general idea of fire performance, an *ad hoc* fire test can be designed specifically for the material's end use. It can sometimes be demonstrated that even though a material does not meet deemed to satisfy requirements it does not present an unacceptable risk. An *ad hoc* fire test is only applicable to one particular situation, so the results of one test, therefore, do not demonstrate an acceptable performance for every situation.

The test to which materials are submitted will depend largely on the planned end use of the material. If a material is to be used to provide blast resistance to a pipe or vessel in the off-shore industry then it could be tested by using explosive charges. If it is required to resist the turbulent attack of a jet fire then the material will be required to demonstrate its resistance in an appropriate fire test. As with all other materials, FRP composites need to have been tested in accordance with the relevant part of BS476.

Selection of composites for fire

In this section the fire performance of different types of polymer composite is discussed, and the treatments for fire protection outlined.

PHENOLICS

Phenolics are generally recognised as having the most inherent resistance to the spread of fire of all polymer composites. Phenolics can achieve the stringent criterion required by London Underground Ltd, in that they can achieve a class '0' fire performance, with nil flame spread being observed in the BS476 part 7 test. Other types of polymer composites can also achieve nil flame spread but require the addition of a fire retardant.

The good fire performance of phenolic composites is due to the high aromatic content present in the matrices. This promotes carbonaceous char formation, which acts as a barrier to:

- the transmission of heat through the material to the undamaged layer beneath the surface causing it to pyrolize, and
- the passage of volatile combustible gases to the surface of the material where they would mix with air and cause flaming combustion.

Although this char formation results in a reduced combustibility and heat release rate, it also has the negative effect of promoting the production of the toxic gas carbon monoxide (CO). CO is a product of incomplete combustion and is a result of the minimal air that is transmitted to the combustion process.

POLYESTERS, VINYL ESTERS AND EPOXIES

All of the above three matrix types do not in general have the same favourable fire performance as phenolics. When tested in accordance with BS476 Part 7, ignition of the matrices is observed almost immediately and continuous flame spread occurs throughout the test. Epoxy and polyester based matrices tend to achieve a class 2 (Up to 455 mm total flame spread in 10 minutes) fire performance.

FIRE RETARDENT ADDITIVES

The fire performance of polyesters, vinyl esters and epoxies can be greatly improved, up to class 0 standard, with the addition of fire a retardant. The most commonly used fire retardant additives are alumina trihydrate (ATH) Filler, phosphorous compounds, zinc compounds, antimony compounds and halogenated compounds based on chlorine or bromine. All of the additives specified interfere with the chemistry of the combustion reaction. This has the benefit of:

- removing heat from the reaction,
- reducing the concentration of combustible gases in the combustion region,
- inhibiting the flow of oxygen to the combustion region.

The fire-retardant material contained within the matrix can affect its structural and weathering performance and therefore the use of phenolic matrices is more widespread where a highly stressed composite is required. If this is likely to be critical then it is particularly important to consult with experienced experts.

THERMOPLASTICS

Thermoplastic FRP composites do not perform well in a fire. They soften and burn and can be a method of spreading fire. The addition of a retardant can affect the fire reaction properties, but the method of retardation must also inhibit the spread of molten burning plastic, and so have the effect of changing the thermoplastic into a thermoset. Therefore the use of thermoplastic materials needs to be planned carefully to avoid significant risks from fire.

SMOKE AND TOXICITY

Some polymer composites produce above average concentrations of smoke when involved in a fully developed fire. The polyester, vinyl ester and epoxy-based matrices tend not to perform as well as phenolic and ATH retardant-based systems, in that the volumes of smoke produced are significantly higher.

The toxic potential of the smoke from fires involving FRP composites is also a consideration that requires investigation. Although many different toxic gases are produced in fire, the main concern is in the production of CO because it is toxic and odourless. If they are forced to burn, phenolics produce more CO that any other polymer composite and ATH-based matrices perform better than phenolics in this respect. However, when tested in accordance with the NES test 713 the overall toxicity index of the phenolics is at a comparable level as epoxy, polyester and vinyl ester matrices. Again the ATH-based matrices perform better than other matrices.

EFFECTS OF ACHIEVING THE FIRE PERFORMANCE ON COMPOSITES

Although the fire performance of the phenolic-based FRP composites is obviously an important advantage from a fire perspective in the selection of an appropriate material, there is the aesthetic disadvantage of its finish. Phenolic FRP composites naturally have a dull red/brown finish, which is not appropriate for many situations. Other FRP composites have the advantage of being able to achieve an aesthetically pleasing finish. Epoxy, vinyl ester and polyester FRP composites are used where a high-quality finish is required. An example of such a situation is the use of GRP in exhibition pieces, which can be very large and so may be required to achieve a level of fire performance.

To achieve a high-quality gloss finish on the surface of the FRP composites, gel coats are used. These gel coats can have a significant detrimental effect on the fire performance of the FRP composite, although gel coats are available that meet Class 1 and Class 2 requirements.

When using polymer composites, it is necessary to consider the fire performance in the early stages of the design process. It is possible to achieve all fire-performance requirements with most FRP composites, provided that they are considered and designed for. The type of FRP composite selected will depend largely upon its required performance in end use. For example, for constructing the interior of underground trains, a phenolic would be the obvious choice whereas, when constructing a large exhibition feature, the use of an epoxy-based matrice would be appropriate. The epoxy-based matrices should be able to achieve a good enough fire performance if designed appropriately.

PRACTICAL FIRE PROTECTION USES OF POLYMER COMPOSITES
Polymer composites have a wide range of fire protection uses, they are used in fire walls, blast walls and blast enclosures. The blast protection usage is common in the offshore industry by virtue of its low weight, low maintenance and easy installation. Blast protection is required in the offshore industry to protect all kinds of equipment, such as pipes that are required to continue transmitting water to a spray deluge system, even if impacted upon by a missile from a nearby explosion or by jet fire impingement. The resulting blast/fire protection enclosure is more efficient than the traditional steel enclosures due to the low thermal conductivity of the polymer and its light weight.

5.2.4   Health and safety

There are a number of health and safety issues unique to FRP composites. It is important that the designer take a balanced view of these concerns, not under-estimating but equally not over-estimating the risks. FRP composites are generally stable, inert materials and are harmless unless they are broken into their constituents. The circumstances under which this may occur can be readily anticipated, and hence appropriate measures can be taken to reduce the risks.

Health and safety considerations for specifying and using FRP composites need to be assessed and managed in a systematic way. The hierarchy of prevention and protection will apply, where hazards must first be eliminated where possible or, failing that, the risk arising from the hazard must be reduced. Once the risk has been reduced as far as practicable, the residual risk must be controlled, firstly with measures that protect everybody and lastly with measures that protect specific individuals (eg personal protective equipment).

Both the resins and fibres of FRP composites can pose a hazard. The uncured resins can be harmful to health through inhalation of vapours and skin contact. Inhalation can lead to headaches, nausea, and loss of consciousness; repeated exposure may lead to damage of internal organs. In many cases contact with the skin can cause dermatitis. However, the fully cured resins are inert and are not a hazard. All types of fibre can cause irritation to the skin, eyes and the respiratory system. There has been concern expressed recently (NCE 2000) about the effect of physical damage on CFRP, resulting in airborne carbon fibres that could affect health. Therefore, extra care is needed when investigating damaged structures. There is no evidence that aramid and glass fibres pose a similar threat, but it is advisable to be cautious and avoid exposure.

Hazards will be present during manufacture, on site if modifications to the FRP composite components are required during installation, in use (particularly if there is a fire), and during demolition. All of these should be considered by the designer under the Construction (Design and Management) Regulations (CDM), and by the contractor under CDM and the Control of Substances Hazardous to Health Regulations (COSHH).

It is particularly important that where FRP composites have been specifically chosen to avoid hazards from manual handling (because of their low weight), this advantage is not negated through insufficient attention to possible hazards posed by the fibres or resins.

Manufacturing

Many of the above health and safety concerns are associated with manufacture and hence are not relevant to this report. It is sufficient to say that manufacturers must ensure that the factory processes safeguard against possible risks. However, if any FRP composite components are produced on site then the rules for the factory production must be applied. This may cause some problems for normal construction practice, in particular, control over ventilation and contamination. A chemical can only cause damage to health if there is exposure, so by preventing contact and exposure to hazardous materials it is possible to control and minimise the risks to health. Knowledge of the chemicals and fibres involved enables suitable training, ventilation, and personal protective equipment to be provided in accordance with the principles of prevention and protection.

Site works

In most circumstances FRP composite components are prefabricated. The individual component may have to be joined *in situ* using adhesives. Again the issues of vapour inhalation and contact with the skin must be addressed. If the hazard cannot be eliminated by selecting safer adhesives, then the risk can be reduced by designing fewer joints. The last resort is to protect against exposure. Generally components will be delivered to size but if they have to be modified by cutting, fibres will be released into the atmosphere. Hence, appropriate steps must be taken to reduce the amount of dust produced and to protect all workers in the vicinity. All dust produced must be gathered and disposed of in the fashion described below.

On the positive side, the low weight of many FRP composite materials as compared to the alternatives, means that some lifting and manual handling hazards can be avoided. These are particularly important in confined spaces, or difficult locations where weight is very important (for example see Section 3.5.2). The risk of fire on site and the hazards from burning FRP composites also need to be addressed. This is covered in greater detail in 5.2.3.

In use risks

Under normal conditions during the service life of the structure the FRP composites should remain stable and hence should not pose a health risk.

If the structure is subject to a fire the FRP composite components will burn releasing toxic gases such as carbon monoxide and hydrogen cyanide (refer to the previous section on fire). With the appropriate fire engineering systems in place this should not pose a threat. The break-down of the matrix will leave the fibres exposed. Where there is no induced stress the fibres will remain in place. However, one should always assume that the fibres would be disturbed because the fire services typically will use high-powered hoses. CFRP can leave sharp shards that do not biodegrade. This can cause traumatic dermatitis on unprotected skin and is considered to be a serious health risk if inhaled or ingested. Such a problem does not arise with glass fibre because the glass melts and hence is no longer fibrous. Aramid fibres tend to survive fires but as they are not rigid they do not pose the dermatitis threat of carbon fibre. However, as mentioned earlier, inhalation of the fibre is to be avoided because the long-term effects are unknown. The fibres do not stay airborne for extended periods of time; therefore it is relatively easy to clean them up. The standard procedure for these materials at an aircraft crash site is to double bag and dispose of them in a low-toxicity waste dump.

There is a potential problem with CFRP where impact damage occurs to a structure, since this may lead to the release of small fibrous particles into the air. There is experience of these problems and how they can be managed from examples of vehicle and aeroplane crashes. The same problems could occur when a structure is being modified, an area may be cut or drilled potentially releasing fibres into the air.

End of life

Decommissioning of the structure should be given some thought. Generally FRP composites cannot be easily recycled. The components should not be demolished on site, and large units may become structurally unstable if they are cut up indiscriminately. It is recommended to take them apart in the way that was intended by the original designer, and then dispose of the pieces according to the guidelines set down by the local authority and the government.

5.2.5   Environmental impact

Introduction

This section explores issues surrounding the use of polymer composites on the environment. As a construction material, polymer composites offer both distinct advantages (mainly in the area of improved performance during the lifetime of the structure) and disadvantages (mainly related to use of chemicals and difficulties with end of life). This document can only give an indicative argument about environmental performance and raise the issues that ought to be considered when using environmental design criteria. The best environmental option will always be a compromise between material groups because they carry out the same functions in different ways.

The most thorough method for examining the environmental aspects of a material is through life-cycle analysis (LCA). In this process (for which ISO 14040 has been developed) the material and its application are studied in a very specific way. Given the wide variety of applications to which FRP composites can be used in construction, it is clearly not possible to carry out a full LCA for all of them. However the approach is valid, as it ensures that all aspects of the life of a material are considered. Here the environmental effects associated with the different stages of the life-cycle are presented, but no attempts have been made to quantify numerically the impacts associated with FRP composites. Reference may be made to Croner (1990).

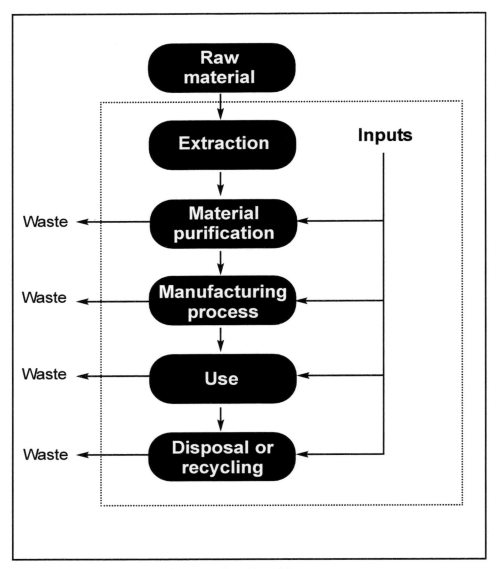

Figure 5.11: *Life-cycle analysis flow chart (Buro Happold)*

Raw materials: resource depletion

The resources used to produce construction materials are often overlooked when assessing the impact of construction. The impact of materials extraction is often valued in terms of energy used; this is usually low in comparison to energy used in the lifetime of a building, but may be significant for other construction types. Other impacts associated with materials extraction include; incidental pollution (in the case of oil), noise, smells, depletion of non-renewable resources and destruction of natural habitats.

All of the matrices and aramid fibres are polymeric materials and hence the environmental issues relating to their raw materials are similar. Polymers form a small part of the use of crude oil, largely being manufactured from the fractions of crude that are not required by the transport industries. Polymers like petroleum jelly Vaseline ®, are by-products of a much larger industry. It can be argued that in some sense, their raw material is a waste product. If polymeric materials are not made from the unused fractions of crude oil, it is likely to be burnt off; this provides an argument for the wider use of polymer-based composites in construction.

Silica glasses use large amounts of non-renewable raw materials, and although they can include recycled material, this is not used in GRP manufacturing. The raw materials for glass manufacture are easily found in the topography of most European nations. The major constituents are silica, soda ash, dolomite and limestone, with traces of feldspar and other natural stone to affect the colour and exact strength of the glass.

Carbon fibres are produced from pitch or polycrylonitrile (PAN), which are not significant in terms of resource use, pitch, in particular, being another by-product.

Manufacturing Processes

RESINS

The manufacture of polymer composites is a carefully controlled process. The raw materials for polymer resin production are (in some cases) dangerous acids that if released into the environment would cause harm, so in all cases particular care must be taken to avoid spillage incidents. The resins are produced in conditions where careful attention is paid to the waste products and environmental hazards, so in most cases there are no major pollutants emitted from these processes. However, where there is a by-product that causes environmental harm it has to be cited with a description of the harm that may result. All of the processes use some energy, but they are not energy intensive in the way that the smelting of metals tends to be.

Epoxy Resins

While the production of epoxy resins does not result in the release of any volatile compounds, caustic agents are used in the processing, which by their nature are toxic to mammals and fish.

Phenolic Resins

Phenolic resins are the result of two types of reaction involving different proportions of phenol and formaldehyde. The Novolac method is the safest of the two and does not result in the release of any volatile compounds. The second method can result in the release of formaldehyde, which is toxic, and when released into the atmosphere it is easily converted into formic acid. Moreover, there are concerns over some of the plasticisers used to make the processing of these resins easier, because they have been linked to hormonal changes in fish.

FIBRES

Glass fibres

The main outputs from the glass manufacturing process are NOx and SOx released from the furnace, and the energy used there. During the process there is a lot of dust from the raw materials as they are moved, but steps are taken within manufacturing plants to reduce levels of dust.

Aramid (Kevlar ®)

Aramid is an aromatic polyamide and it is often copolymerised with Nylon 6, which increases the flexibility of the fibres and improves wear. This process has no major pollutants. Steam is a by-product of the polymerisation process, but this is usually captured and recycled into the re-hydrolysis of the fibres further on in the process.

Carbon fibres

Carbon fibres are produced from the pyrolysis of pitch or polyacrylonitrile (PAN) fibres or, less frequently cellulose; they are subsequently heat treated. At low temperatures the gases N_2, H_2 and HCN are formed, but the latter two do not survive the elevated temperatures of the process and are not released into the atmosphere.

Transport

The transport of the materials both before and after processing can represent a significant environmental impact for some materials, particularly the heavy aggregate type materials. Clearly FRP composites are not in this category, and generally the low weight of FRP composites will represent a benefit in terms of reduced transportation energy and other impacts.

Use/lifetime

One of the principle benefits of using polymer composites is that it can increase the lifetime of structures in corrosive environments, be it through their original design or subsequent repair with FRP composite materials. This means that the environmental impacts discussed in this chapter will generally need to take into account the avoided impacts from the required retrofit and maintenance of structures.

Furthermore, some FRP composite systems can contribute to the energy efficiency of buildings, on account of their inherent insulation properties. This means that less insulation is needed for cladding systems incorporating FRP composites than for metal alternatives. The same applies for window or door systems, and should be taken into account in the design of the building; lower in-life energy use should be possible.

Disposal and end-of-life

Re-use

As a consequence of the bespoke nature of construction products unless a very transferable item is manufactured there is little opportunity for components to be re-used without significant modification. Although there are no current plans to reuse composites, good labelling of constituent parts would allow for such decisions to be made in the future.

Recycling

Recycling of polymers is carried out more often than it has been in the past. However a composite material is far more difficult to recycle than a single material, precisely because of its composite nature; it is very difficult to separate the resin from the fibres, and therefore very difficult to recycle the materials into more of the same product. This means that any recycling that is possible will result in a lower quality end product.

Primary recycling is direct re-use. Secondary methods result in the material remaining the same chemically but they are structurally altered. Tertiary methods change the chemical nature of the material. There is a greater input of energy and resources in proportion to the amount that the material is altered.

As re-use is rare there are two methods for recycling composites:

1 A secondary method that involves the mechanical destruction of the articles into small chips that can be used as a filler for new composites.

2 A tertiary method that breaks down the composite into its constituent parts by chemical and thermal decomposition.

The first step in the recycling of any polymer-based composite is the deconstruction into small-sized pieces. This in itself presents a problem, because of the toughness of the material (the very reason for its use), shredding and chipping is a very expensive and time-consuming process and it is also energy intensive.

Studies have shown that up to 15 per cent scrap filler can be added to polyester based composites without a deterioration of the materials' properties. The main issue with this method of recycling is that the resultant composite will not possess the same texture as one that is made from virgin materials and often needs a gel coat of virgin materials to give the required finish.

Tertiary deconstruction of polymer composites is possible in a variety of ways, the most useful of which are hydrolysis and pyrolysis. Hydrolysis involves the addition of water at elevated temperatures under normal pressure in a controlled environment. The results are low weight monomers, oligomers and clean fillers. Pyrolysis involves the heating of the material in the absence of oxygen, the resulting materials are charred organic compounds and fillers. This method is available to a wider range of composites than hydrolysis and it leaves the filler clean enough for reprocessing.

Polymeric articles are currently labelled for the easy identification of the compound for recycling. This has worked well in the motor industry and led to significant levels of recycling. Similar levels of standardisation for the labelling of composite materials could be adopted by the construction industry. However, it is possible that if a composite was separated into its constituent parts the polymeric element might not be one that was identical in every way to feed stock for other applications. An example of this is that polyethylene (PE) bottles are recycled, but they may be shredded and placed with PE carrier bags, where the structure of the PE is slightly different, the resultant blended feed stock would not necessarily be useful for either application.

Whether polymeric composites will ever be recycled as readily as other polymers or metals is mainly an economic issue. If the value of the fillers increases, or the demands for thermoset feedstock can no longer be supplied by the oil industry the balance will be pushed further towards recycling. Until this is the case wholesale recycling of composites is unlikely.

Conclusions

The designer needs to compare materials through the analysis of the overall environmental impact (whether negative or positive) of using FRP composites rather than traditional materials. At the heart of this debate will be the question of whether the high ratio of strength (modulus) to embodied energy that makes some composites an alternative to steel (Ashby 1999) is capable of balancing the end of life environmental problems associated with composites.

A full life-cycle analysis (LCA) is not needed for every item in every project, but each feature should be given consideration, and a balanced conclusion should be reached. These choices are not straightforward as all materials bring with them environmental impacts that need to be accounted for. In general FRP composites will be lighter and use less energy than competing products, but there are different pollution issues, and a problem with end-of-life/recycling as compared to metals.

5.2.6 Finishes and aesthetics

One of the great advantages of FRP composites is the almost limitless range of finishes and appearances that can be achieved. The mould and the gel coat can be adjusted to achieve virtually any desired surface finish.

When designing in FRP composites, the designer is free to choose virtually any appearance and texture, which may be required, and then work with the manufacturers to produce it. Most manufacturers will be able to provide samples and brochures to give the designer some idea of the possibilities, and the examples shown in this report may provide further inspiration.

The range of finishes available to the designer deserves further investigation. In addition to simply varying the colour, the designer may select a matt, gloss or metallic finish, or a patterned finish with coloured or metallic flecks. The surface texture may be smooth, or textured to imitate natural materials like stone or timber. The surface can incorporate elements such as a non-slip finish for trafficable areas, and elements such as lights and reflective strips can also be included. Particular applications of this are the reproduction of and matching to "real" materials like stone, that can be produced at much lower weight than the originals.

Examples of this include the work at Liverpool Cathedral (Section 3.3.6), the Rest Zone in the Millennium Dome (Section 3.5.3), Wiltshire Radio Station (Section 3.1.10) and the images in Section 4.2.1, which show the range of products available.

IS THE APPEARANCE OF FRP COMPOSITES A POSITIVE OR A NEGATIVE FEATURE?
There is a wide range of perceptions regarding the aesthetics of FRP composites. Some may view the ability to copy natural materials as a great benefit. An example of this occurs in the repair of historic buildings, where cheaper and lighter FRP composites may be used to replace ornate stonework. Others may see the reproduction of a natural material in such a situation as inappropriate and undesirable. Similarly there is an image associated with performance racing cars, which some see as positive, but plastic cars do not carry the same image. There are no right and wrong answers to these speculations, other than to say that each designer must decide, which material is appropriate in each individual situation.

The issue behind these different perceptions is really just a reflection of the versatility of the materials. While they are perceived very differently, Formula One cars and Robin Reliants both use FRP composites as the most appropriate materials for the car bodies. It is not for this guide to enter into the debate about perceptions of aesthetic properties of FRP composites, but rather to ensure that the designer considers the full range of options available when designing a product.

POTENTIAL PROBLEMS WITH APPEARANCE

Older samples of FRP composites show a variety of problems, some of these are discussed in Section 5.2.2. The three main issues are colour fading, crazing and chalking. In more recent products all of these have been addressed through improvements in the make up of gel coats in particular.

1   Colour fading: Darker colours are more prone to fading as they absorb more of the UV light that causes the colour loss. Hence avoiding darker colours on external surfaces, particularly south facing, will help. Also a design that will not suffer visually through the variable nature of colour fading (not all surfaces receive as much light) may help in this regard.

2   Crazing: Crazing is the appearance of fine hair cracks on the resin surface of an FRP composite, and it may develop immediately after manufacture or after some months of use. Its presence may be indicated initially simply as a loss of surface gloss. The phenomenon is associated with the presence of resin-rich areas, and may occur as a result of the use of an unsuitable resin or inappropriate resin formulation. For example, if a thick gel coat is being used, crazing or cracking may occur unless a flexible resin formulation is chosen. Crazing in service may be a result of weathering or chemical attack, resulting from under-cure of the resin, the use of too much filler, or the use of a resin that is too flexible.

3   Chalking: Chalking is a result of the migration of filler particles to the surface, as a result of exposure to moisture and is exacerbated by under-curing of the resin. Chalking occurs more rapidly the higher the temperature becomes. The use of appropriate filler surface treatments and resin formulation help to limit this problem. Undercuring can also lead to leaching of the resin itself as a result of exposure, and this is much more dangerous than chalking because it leaves the reinforcing filaments exposed to attack by moisture.

These and other faults in GRP laminates, together with suggested remedies, are discussed in greater detail by Scott Bader (1994).

PAINTING

Although many reinforced plastics laminates are used in the as moulded condition, and rely on the gel coat and pigmenting to provide the necessary colour, texture and durability, there are many applications for which painting is an appropriate finishing operation. Car body mouldings, for example, are often sprayed and stoved to provide the conventional Class A automotive finish, and aircraft component surfaces are frequently painted with pigmented epoxide paints, both for aesthetic reasons and also to limit the ingress of environmental moisture into the internal composite structure. If painting is to be carried out, it is important to ensure that all traces of mould-release agents on the laminate surface are removed either mechanically or by means of solvents prior to painting, and that appropriate primers with good adhesion to polymer surfaces are used. Air-drying paint finishes can usually be applied without post-curing of the resin, but if an article is to be stoved or if cellulose paints are to be used it is necessary to ensure that the resin is fully post-cured.

A clear sign of the performance that GRP materials can achieve is shown by the fact that they are able to meet the stringent "Class A" finish specified for automobile bodies. This means that it is not possible to detect by eye that parts of some cars are in fact made of GRP instead of metal. There are a number of visual criteria that may be specified for the materials. These include:

1   Reflectivity and dazzle. This defines the amount of reflected light, and the amount of Gloss can be specified according to BS 2782 Part 5.

2   Colour. Any colour can be achieved with pigments to change the basic colours of the resins. Typical colour ranges for polyesters are defined in the BS4800 range. The light fastness defines the length of time a colour stays unfaded, and can be specified with BS1006. For external use pigments should be able to achieve at least 6 on the light fastness scale.

3   Texture. Materials can be given a range of textures, by changing the surface of the mould tool, from smooth surfaces to those similar to brick or concrete.

6 Procurement

There are some FRP composite materials that are available to be sold as standard components, for example pultruded beams of certain dimensions, or CFRP plates for beam strengthening. In these cases the procurement is fairly straightforward, once the appropriate manufacturers have been located.

Therefore this section concerns the procurement of non-standard, purpose-designed moulded units, typically for buildings, but with other possible applications. The units may be cladding or other external or internal decorative components, or they may have a more important structural requirement.

The procurement process may be modified by the overall procurement mode that has been adopted for the building. This may be a traditional process through a main contractor or at the other extreme a "Construction Management" route where all the work is let as separate packages. Whatever route is adopted, the units will be supplied through a specialist contractor who will have considerable skill and experience in detailing and making FRP composite units. The aim of the process is to set up a framework in which the skills and knowledge of the contractor can be utilised to optimise the design and get the best quality while cost control can be maintained. In the case of procurement routes other than construction management this may involve adapting the main contract process to include more input from the package contractor.

It is particularly important to obtain input from specialists because of the absence of codes and standards for most types of FRP composites. This means that the traditional procurement routes are more difficult to use for projects involving FRP composites.

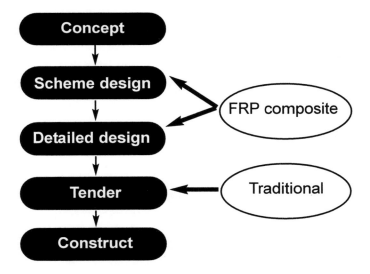

Figure 6.1: *Flow diagram for procurement, showing traditional stage for contractor input, and that needed for FRP composite projects*

Figure 6.1 shows the main stages of a typical construction project. In a traditional project the contractors only get involved at the tender stage, when they are asked to compete for the chance to build what has been designed. Recently there has been a great deal of discussion and demonstration of the benefits of involving contractors and specialist or trade contractors at earlier stages in the process. This is particularly valuable in projects involving FRP composites, where the specialist contractors and materials suppliers have the best knowledge of their own materials, and how they work.

Not all FRP composite suppliers and contractors have design capability, and it will often be important to consider a range of companies that offer different technologies. Therefore, it can be worth considering employing a specialist consultant to aid the design of the FRP composite part of the construction. This is quite common for even the most well-established consulting engineers, who cannot be experts in every aspect of every material.

Concept

This report aims to make designers sufficiently aware of the capabilities of FRP composites, so that they will be considered at concept design. The many issues raised throughout this report, but particularly performance and whole-life costs should be considered. From this stage the possibility of using FRP composites should appear as an option for the next stage.

Scheme design

It is in the scheme design stage that the first opportunity for specialist input should arise. If the FRP composite options are looking possible, then the design team will pursue these further, and investigate to a certain extent the wide range of technical issues discussed throughout this report. An important point to check at this stage is whether the programme for construction is going to be helped or hindered by the FRP composite option. There can be gains from faster installation of lighter products, but some parts may have long lead times because of manufacturing constraints and the need for tooling.

Emerging from the scheme design should be a "half worked" design that would clearly show where and how the FRP composites are proposed to be used. Alternatives may well be retained in case of subsequent problems. One risk at this stage is that quantity surveyors will be less familiar with the materials, and may therefore put relatively high cost estimates on an FRP composite option.

Detailed design
During detailed design, the design team produce a full set of design drawings, backed up by calculations and consultations with others. The structural engineer for the project should be able to perform basic calculations to size the components using typical data for strength and stiffness of the composite. This should be done alongside the architectural definition of the form and other requirements. For special components where stiffness or strength is critical and particularly if carbon or aramid fibre is being used the responsibility for designing a composite to meet those requirements should be passed to the contractor in the specification. He/she may need to employ a specialist engineer to carry out the calculations and define the detailed lay-up requirements.

Clearly this process involves detailed work and costings, and always an element of iteration and design development. For unusual products, or appearance critical pieces, it will help if test samples are produced. These can demonstrate the performance in structural terms, or for colour matching.

Tender

By the tender stage the design should be fixed, and only relatively minor changes are likely to be possible. It is for this reason that input from FRP composite specialists should be obtained as early as possible – this stage will generally be too late to change the whole solution.

Unusual geometries

We have noted that it is a virtue of FRP composites that a very wide range of geometries can be produced. If the component is to be made to a doubly curved geometry then this will normally be generated and analysed by the use of computer software. This work of processing the geometry is also a specialist task, which may be done by the engineer or by a specialist designer, perhaps with the input from a mould-making company. The development of the forms should be sent to the design team for approval.

Support for the making of moulds can be found from specialists. The Gauge and Toolmakers Association can provide contacts for this – see Section 7.1 for the address.

Quality control

Structurally the performance of the unit depends on the amount and arrangement of the fibres in the composite and this cannot easily be checked after completion. The lay-up will normally have been defined by the contractor's engineer but there may be a need to check that the correct materials have been used and that it has been correctly laid up. This would require independent inspection, perhaps by the designer. For some units such as a beam or a mast, the stiffness can be checked by measuring the deflections in a load test.

Durability is affected by the workmanship as well as by the selection of suitable materials. It is particularly important to have the correct thickness of gel coat and thorough wetting of the fibres by the resin. The density of the resin can be improved by removing the entrained air by vacuum bagging, if this is possible, within the mould configuration. These topics were addressed in detail in Section 5.2.2.

The building owner would be looking for a life of cladding units of say 50 years. Over this period guarantees from fabricators may not have much value and any responsibilities of the design team would not extend beyond 12 years. The owner's best protection is to select a contractor with a good track record, use the best available materials and if necessary employ an inspector.

It may help to look for designers and suppliers with appropriate certification, to ISO 9000, 9001 or 9002 accordingly.

This report does not attempt to deal with the detail of specification documentation. It will be important to set these out in terms of materials, workmanship and tolerances, in order to ensure that the right product is produced for the job. It may also be necessary to have a number of samples prepared, to check that the tolerances and specification can be met. The cost impact of this may not be small, especially if tooling is required prior to manufacturing, and so only reasonable requests should be made.

7   Appendices

7.1   APPENDIX A: SOURCES OF FURTHER INFORMATION

7.1.1   Directories

Applegate Plastics and Rubber Directory,
www.applegate.co.uk/index2.html

British Plastics and Rubber on-line directory,
www.polymer-age.co.uk/main.htm

7.1.2   Research organisations and professional institutes

Advanced Composites Manufacturing Centre (ACMC),
Department of Mechanical and Marine Engineering,
University of Plymouth, Drake Circus, Plymouth, PL4 8AA
Tel: +44 (0) 1752 232651
www.tech.plym.ac.uk/sme/acmc.htm

AEA Technology,
Harwell,
Didcot, Oxon.
www.aeat.co.uk/

Building Research Establishment (BRE),
Bucknall's Lane,
Garston,
Watford WD2 7JR, Herts
Tel: +44(0) 1923 664000
www.bre.co.uk

British Plastics Federation (BPF),
6 Bath Place,
Rivington Street,
London EC2A 3JE
Tel: +44 (0) 20 7457 5000
www.bpf.co.uk/index.htm

Centre for Advanced Composites in Construction (CACIC)
Universities of Southampton and Surrey
Professor L Hollaway,
Department of Civil Engineering,
University of Surrey;
Guildford, Surrey, GU2 7XH,
Tel: +44(0) 1483 879532.

Centre for Composite Materials Engineering
Professor AG Gibson,
Herschel Building, Materials Division,
Department of Mechanical, Materials and Manufacturing Engineering,
University of Newcastle upon Tyne NE1 7RU
Tel: +44 (0) 191 222 6173
www.ncl.ac.uk/materials/materials/resgrps/CCME.html

Composite Materials Centre
Professor F Matthews,
Imperial College,
Prince Consort Road
South Kensington, London, SW7 2AZ
Tel: +44 (0) 20 7589 5111
www.cm.ic.ac.uk

Composite Materials Research Group
Professor P A Smith,
School of Mechanical and Materials Engineering
University of Surrey
Guildford, Surrey GU2 7XH
www.surrey.ac.uk/MME/Research/research_composite.html

Composites Processing Association,
K. Fosdyke, Sarum Lodge,
St, Anne's Court, Talygarn, Pontyclun, CF72 9HH
Tel: +44(0) 1443 228867
www.composites-proc-assoc.co.uk

Concrete Society,
Telford Avenue
Crowthorne RG45 6YS, Berkshire
Tel: +44(0) 1344 466007

Composite Structures Research Unit,
Professor L Hollaway,
Department of Civil Engineering,
University of Surrey;
Guildford, Surrey, GU2 7XH,
Tel: +44(0) 1483 879532,
www.surrey.ac.uk/CivEng/research/csru/index.htm

Defence Evaluation and Research Agency (DERA),
Ively Road,
Farnborough
Hampshire GU14 0LX
Tel: +44 (0) 1252 393300
www.dra.hmg.gb/index.php3

Gauge and Toolmakers Association
3 Forge House
Summerleys Road
Princes Risborough
Bucks
HP27 9DT
01844 274222
www.gtma.co.uk

Institute of Materials
1 Carlton House Terrace,
London, SW1Y 5DB
Tel: +44 (0) 20 7451 7300
www.materials.org.uk/

Institution of Mechanical Engineers,
1, Birdcage Walk,
London SW1H 9JJ
Tel: +44(0) 20 7222 7899
www.imeche.org.uk/

Institute of Polymer Technology
and Materials Engineering
Loughborough University.
Tel: +44 (0) 1509 223331
www.lboro.ac.uk/departments/iptme/

Lightweight and Composite Structures Group
Professor RA Shenoi
School of Engineering Sciences.
University of Southampton
Southampton SO17 1BJ
Tel: 44 (0) 23 8059 2356
www.soton.ac.uk/~shipsci

Network Group on Composites in Construction (NGCC),
President, Dr S Halliwell
Building Research Establishment,
Tel: +44(0) 1923 664834
www.ngcc.org.uk/

National Physical Laboratory,
Teddington, Middlesex,
UK, TW11 0LW
Tel: +44(0) 20 8977 3222
www.npl.co.uk/

Polymer Composites Group
Professor C D Rudd
M3EM, University of Nottingham,
University Park,
Nottingham, NG7 2RD. UK.
Tel: +44 (0) 115 9513770
www.nottingham.ac.uk/~eazwww/composite/

Production Engineering Research Association (PERA)
Technology Centre
Melton Mowbray
Leicestershire LE13 0PB
Tel: +44 (0) 1664 501501
www.pera.com/composites/composites.html

Royal Institute of British Architects Client Services
66 Portland Place
London W1B 1AD
RIBA Members Information Line: +44(0) 906 302 0444
Public Information Line: +44(0) 906 302 0400
site.yahoo.net/client-services/

Rubber and Plastics Research Association,
Shawbury, Shropshire
Tel: +44 (0) 1939 250383
www.rapra.net/home.html

Society for the Advancement of Material and Process Engineering (SAMPE)
International Business Office
1161 Parkview Drive
Covina, California, 91724-3748, USA
Tel: +1 626-331-0616
www.sampe.org/

Structures and Materials Group
Dr GJ Turvey
Department of Engineering ,
University of Lancaster
www.comp.lancs.ac.uk/engineering/research/structures/fibre.html

The Welding Institute (TWI),
Granta Park,
Great Abington,
Cambridge CB1 6AL
Tel: +44 (0) 1223 891162
www.twi.co.uk

7.1.3 Information locations

Advanced Composites Applications in the UK
Website maintained by Dr CJ Burgoyne,
Department of Engineering, University of Cambridge,
Trumpington Street, Cambridge.
www-civ.eng.cam.ac.uk/isegroup/uklocate.htm

Composite Basics
www.cmi-composites.com/basics.htm

Composites Link
www.compositeslink.com

Composites On-Line data base and URL listings
www.composites-online.com

Infrastructure Applications of Composite Materials,
iti.acns.nwu.edu/projects/comp.html

NetComposites (on-line data base),
Mr G Bishop,
Treak Technology Ltd, PO Box 9898,
Melton Mowbray,
Leicestershire, LE14 3ZH
Tel: +44(0) 941 119898
www.netcomposities.com

Research Focus
www.ice.org.uk/public/rf29_p13.html

7.2 APPENDIX B: KNOWN EXAMPLES

In addition to the examples referred to in the case studies and in the applications chapter, there is a large number of other project examples that have been built. A number of these are listed below, but the list is by no means exhaustive.

Table 7.1: *A selection of known examples*

Name of project	Country	Location	Date	Material
Buildings				
Web of life building	UK	London	1999	GRP
BRE Building 4	UK	Watford	1987	
Broadmoor Hospital, room concept	UK	Crowthorne	1987	
Tredegar Street car park	UK	Cardiff	1985	
Blackpool ambulance station	UK	Blackpool	1984	
Amoco building	UK	Swindon	1983	GRP
Factory unit, Nottingham	UK	Nottingham	1980	
Sports hall IBM	UK		1980	
Corner units: Winwick Quay	UK	Warrington	1978	
Factory at Winwick Quay	UK	Warrington	1978	GRP
Ballincollig community school	Ireland		1976	GRP
French music publishers			1975	
Mondial House, cladding	UK	London	early 1970s	GRP
HM Prison Long Marton, wall cappings	UK	Long Marston	1970s	GRP
Service Tower Sussex Gardens	UK	London	1967	
Bali Stadium	Bali, S Pacific			
Raybould House	Canada	Toronto		
Swimming Pool, Douvrin	France	Douvrin		GRP
Kings College Hospital	UK	London		
Legoland	UK	Windsor		
River Police Building, Wapping	UK	London		
Stratford Market depot	UK	London		
Waterloo Station, roof glazing bar connections	UK	London		

Continued on next page

Name of project	Country	Location	Date	Material
Westway shell roofs	UK	London		
Experimental building	USA	W Virginia		
L A Ontario airport	USA	Ontario		
Niko hotel, Beverley Hills	USA	Los Angeles		GRP
Bridges				
Oppegard golf company	Norway		1997	GRP+Kevlar
Headingley bridge	Canada		pre-1997	CFRP/GFRP
Mukai Bashi bridge	Japan		pre-1997	CFRP cables
Storchen bridge	Switzerland		pre-1997	CFRP cable
Mugen bridge	Japan		1996	AFRP flat rod
Second Severn crossing bridge enclosure	UK	Gloucs	1996	GRP
McKinleyville bridge	USA		1995	GFRP bars
Centre Street, bridge	Candada		pre-1995	CFRP tendons
Tring bridges	UK		1994	Parafil rope
Hakui bridges	Japan		1991/1992	CFRP cable
Kagawa railway	Japan		1991	AFRP
Rapid city bridge	Japan		1991	G+CFRP/steel
Oyama bridge	USA		1990	AFRP
A19 Tees viaduct, bridge enclosure	UK		1989	GRP
Shinmiya bridge	Japan		1988	CFRP cable
Tendon bridge	Germany	Dusseldorf		
Oppegaard footbridge	Norway	Oslo		
Expo footbridges, Lisbon 98	Portugal	Lisbon		
Cornwall bridges	UK	Cornwall		
Wroxham footbridge	UK	Norwich		
Repair/strengthening				
Dintellhaven bridge	Netherlands		1999	CFRP cables
Clearwater creek	Canada		pre-1997	CFRP plate
Canossiane conv.	Italy		pre-1997	Arapree
Yacht club	Thailand		pre-1997	FRP plate
Houston garage	USA	Houston	pre-1997	CFRP plate
Norfolk Navy base	USA	Norfolk	pre-1997	CFRP plate
Champlain bridge	Canada		1996	GFRP wrap
Highway 10	Canada		1996	GFRP/CFRP
Webster garage	Canada		1996	GFRP/CFRP
Oberriet bridge	Switzerland		1996	CFRP plate
Sherbrooke University	Canada		1995	GFRP sheet
Gossau city hall	Switzerland	Gossau	pre-1995	CFRP plate
Uzwil supermarket	Switzerland	Uzwil	pre-1995	CFRP plate
Zurich railway station	Switzerland	Zurich	pre-1995	CFRP plate
Leibstadt power station	Switzerland	Leibstadt	pre-1995	CFRP plate
Nikkureyama tun	Japan		1994	CFRP plate
Shirota bridge	Japan		1994	CFRP plate
Glendale repair	USA		1994	GFRP plate
Flims garage	Switzerland	Flims	1993	CFRP plate
Metro railway	Japan		pre-1993	CFRP plate
Brighton Pavillion, repair works	UK	Brighton		
Midland hotel	UK	Glasgow		
Jack arches	UK	London		

Continued on next page

Name of project	Country	Location	Date	Material
Other				
Millenium Beacon	UK	London	1999/2000	
Home planet	UK	Greenwich	1999	
Meridian Garden statues	UK	Greenwich	1999	
Self portrait	UK	Greenwich	1999	
The body	UK	Greenwich	1999	
Fence in Hemel Hempstead	UK	Hemel Hempstead	1996	FRP bars
Tarmac concrete	UK		1995	GFRP rods
Meishin express	Japan		1994	AFRP tendons
AWRE, clothes racks	UK	Aldermaston	1986	
Channel Tunnel cable trays	UK/France		1980s	GRP
Mass Transit system	Malaysia			
Barcelona Tower, cables	Spain	Barcelona		Polymer tendons
Scrubber Units for Drax	UK	Drax		GRP (VE)
Northumbrian Water, odour control scheme	UK	Howden, Tyne & Wear		

8 References and bibliography

8.1 TEXT REFERENCES

Alderson A, 1999, *Fibre-reinforced polymer composites for blast-resistant cladding,* Proceedings of a conference on Composites and Plastics in Construction, Nov 1999, BRE, Watford, UK (RAPRA Technology, Shawbury, Shrewsbury, UK) paper 22 1–9

Amery C, 1995, *Architecture, Industry and Innovation: The Early Work of Nicholas Grimshaw and Partners* (Phaidon Press, London)

Anon, 1997, The Fibreline bridge: a break-through in composite design technology, *Mater and Design 18*, 43

Ashby M F and Jones D R H, 1980, *Engineering Materials, Vol 1,* (Pergamon Press, Oxford)

Ashby M F, 1999, *Materials Selection in Mechanical Design (Second edition)* (Butterworth-Heinemann, Oxford)

Astrom B T, 1997, *Manufacturing of Polymer Composites* (Chapman and Hall, London)

Aveston J, Kelly A and Sillwood J M, 1980, Long-term strength of glass-reinforced plastics in wet environments, in *Advances in Composite Materials: Proceedings of the 3rd International Conference on Composite Materials,* ICCM3 (Paris), Vol 1, (Pergamon, Oxford) 556–568

Bader M G, Smith W, Isham A B, Rolston J A and Metzner A B, 1990, *Processing and Fabrication Technology: Vol.3 of The Delaware Composites Design Encyclopaedia* (Technomic Publishing Co Inc, Lancaster, Pennsylvania, USA)

Berry D B S, 1990, Fire properties of polymers and composites, *Polymers and Polymer Composites in Construction* (Editor, Hollaway LC) (Thomas Telford Ltd, London) 168–181

Broek D, 1986, *Elementary Engineering Fracture Mechanics* (Martinus Nijhoff, Dordrecht, Netherlands)

Brookes A J, 1998, *Cladding of Buildings* (E and F Spon, London)

BSI, 1987, *Specification for Design and Construction of Vessels and Tanks in Reinforced Plastics:* BS 4994 (British Standards Institute, London)

BSI, 1999, *Reinforced Plastics Composites—Specification for Pultruded Profiles: parts 1,2 and 3* (British Standards Institute, London) prEN 13706–1, prEN 13706–2 and prEN 13706–3

Bunsell A R, 1988, *Fibre Reinforcements for Composite Materials – Composite Materials Series Volume 2,* (Elsevier Amsterdam)

Burgoyne C J and Head P R, 1993, *Aberfeldy Bridge – an advanced textile-reinforced footbridge*, Proceedings of TechTextil Symposium, June 1993, Frankfurt, Germany, paper 418 1–9

Burling P, 1999, *Heat insulation and fire resistance of composite materials*, Proceedings of a conference on Composites and Plastics in Construction, Nov 1999, BRE, Watford, UK (RAPRA Technology, Shawbury, Shrewsbury, UK) paper 25 1–5

Busel J P and Lindsay K, 1997, On the road with John Busel: A look at the world's bridges, *CDA/Composites Design and Application*, Jan/Feb 14–23

Cadei J M C, 1998, Factors of safety in the limit-state design of FRP composites, in *Proceedings Intl Conference on Designing Cost-Effective Structures* (Institute of Mechanical Engineers, London)

Clarke J L and O'Regan P, 1995, The UK's first footbridge reinforced with glass-fibre rods, *Concrete 29 (4)* 31–32

Clarke J L (Editor), 1996, *Structural Design of Polymer Composites: Eurocomp Design Code and Handbook* (E and FN Spon, Chapman and Hall London)

Concrete Society, 2000, *Design Guidance for Strengthening Concrete Structures using Fibre Composite Materials*: Technical Report 55 (Concrete Society, Crowthorne, Berks)

Crane F A A, Charles J A and Furness J, 1997, *Selection and Use of Engineering Materials* (Butterworth-Heinemann, Oxford)

Croner Publications, 1990, *Substances Hazardous to the Environment* (Croner Publications, London)

Davidge R W, 1979, *Mechanical Behaviour of Ceramics* (Cambridge University Press Cambridge)

Dodds N, Gibson A G, Dewhurst D and Davies J M, 2000, Fire behaviour of composite laminates, *Compos A: Appl Sci and Manufg, 31A* 689–702

Donnet J B, Bansal R C, 1990, *Proceedings of the 8th European Conference on Composite Materials* ECCM8, Naples, June 1998, Symposium 3, Volume 2 (Marcel Dekker New York)

Dorey G, 1982, Fracture of composites and damage tolerance, *Practical Considerations of Design Fabrication and Tests for Composite Materials – AGARD Lecture Series no 124* (Director B Harris) (AGARD/NATO Neuilly Paris) paper 6

Dorey G, 1982, Environmental degradation of composites, *Practical Considerations of Design Fabrication and Tests for Composite Materials – AGARD Lecture Series no 124* (Director B Harris) (AGARD/NATO Neuilly Paris) paper 10

Eckold G C, 1994, *Design and Manufacture of Composite Structures* (Woodhead Publishing, Abington, UK)

Fibreline, 1995, *Fibreline Design Manual for Structural Profiles in Composites* (Fibreline Composites A/S, Kolding, Denmark)

Gibson RF, 1994, *Principles of Composite Material Mechanics* (McGraw-Hill, New York)

Gilby J, 1998, Pultrusion provides roof solution, *Reinforced Plastics 42 (6)* 48–52

Goldstein B, 1978, *Progressive Architecture* (Trim-Tech)

Halliwell S M, 1999, *Reinforced plastics cladding panels,* Proceedings of a conference on Composites and Plastics in Construction, Nov 1999, BRE, Watford, UK (RAPRA Technology, Shawbury, Shrewsbury, UK) paper 20 1–5

Halpin J C, 1992, *Primer on Composite Materials Analysis, Second Edition* (Technomic Publishing Co Inc, Lancaster PA, USA) 180–187

Hancox N and Meyer R M, 1994, *Design Data for Reinforced Plastics* (Chapman and Hall, London)

Harris B, Dorey S E, Cooke R G, 1988, Strength and toughness of fibre composites, *Compos Sci and Technol 31* 121–141

Harris B, 1994, Fatigue of glass-fibre composites, in *Handbook of Polymer-Fibre Composites* (Editor Jones F R) (Longman Green, Harlow, UK) 309–316

Harris B, 1994, Non-destructive evaluation of composites, in *Handbook of Polymer-Fibre Composites* (Editor Jones F R) (Longman Green, Harlow, UK) 351–356

Harris B and Phillips M G, 1983, Non-destructive evaluation of the quality and integrity of reinforced plastics, *Developments in GRP Technology-1* (Editor Harris B), (Applied Science Publishers, London), 191–247

Harris B and Phillips M G, 1990, Defects in manufactured composites, *Analysis and Design of Composite Materials and Structures* (Editors Surrel Y, Vautrin A and Verchery G) (Editions Pluralis, Paris) Part 1, Chapter 12

Harris B and Phillips M G, 1990, Non-destructive evaluation of composites, *Analysis and Design of Composite Materials and Structures* (Editors Surrel Y, Vautrin A and Verchery G) (Editions Pluralis, Paris) Part 1, Chapter 15

Harris B and Phillips M G, 1990, Damage in composites in testing and service, *Analysis and Design of Composite Materials and Structures* (Editors Surrel Y, Vautrin A and Verchery G) (Editions Pluralis, Paris) Part 1, Chapter 13

Harris B, 1996, Fatigue behaviour of polymer-based composites and life prediction in *Durability Analysis of Structural Composite Systems* (AA Balkema, Rotterdam) 49–84

Harris B, 1999, *Engineering Composite Materials (second edition)* (Institute of Materials, London)

Hart-Smith L J, 1993, The ten percent rule, *Aerospace Materials 5 (2)* 10–16

Harvey K, Ansell M P, Mettem C J and Bainbridge R J, 2000, *Bonded-in pultrusions for moment-resisting timber connections*, Proceedings of 33rd Meeting of Timber Structures Working Commission, CIB-W18, Delft, The Netherlands, August 2000 (ICRIBC/University of Karlsruhe, Karlsruhe, Germany)

Head P R and Templeman R B, 1990, Application of limit-state design principles to composite structural systems, in *Polymers and Polymer Composites in Construction* (Editor Hollaway L C) (Thomas Telford Ltd, London) 73–93

Head P R, 1994, Limit state design method, in *Handbook of Polymer Composites for Engineers* (Editor Hollaway L C) (Woodhead Publishing, Abington, UK) Chapter 7

Hill P S, Smith S and Barnes F J, 1999, *Use of high-modulus carbon fibres for reinforcement of cast-iron compression struts within London Underground: project details,* Proceedings of a conference on Composites and Plastics in Construction, Nov 1999, BRE, Watford, UK (RAPRA Technology, Shawbury, Shrewsbury, UK) paper 16 1–6

Hinton M J, Soden P and Kaddour A S, (Editors), 1998, Failure Criteria in Fibre-Reinforced Polymer Composites: Special Issue of *Composites Science and Technology 58* 999–1254,

Hogg P J and Hull D, 1983, *Corrosion and environmental deterioration of GRP, Developments in GRP Technology–1* (Editor Harris B) (Elsevier Applied Science, London) 37–90

Holliday L and Robinson J D, 1977, The thermal expansion properties of polymer composites, *Polymer Engineering Composites* (Editor MOW Richardson) (Applied Science Publishers, London) 263–315

Hudson F D Hicks I A and Cripps R M, 1993, The design and development of modern lifeboats, *J Power and Energy 207* (Proc I Mech E; A) 3–12

Hutchinson A R, 1997, *Joining of Fibre-Reinforced Polymer Composites* CIRIA Project Report 46, (CIRIA, London)

Jones F R, Rock J W and Bailey J E, 1983, The environmental stress-corrosion cracking of glass-fibre-reinforced laminates and single E-glass filaments, *J Mater Sci 18* 1059–1071

Jones F R, Rock J W and Wheatley A R, 1983, Stress-corrosion cracking and its implications for the long-term durability of E-glass fibre composites, *Composites 14* 262–269

Jones F R, 1999, Durability of reinforced plastics in liquid environments, in *Reinforced Plastics Durability* (Editor Pritchard G) (Woodhead Publishing, Abington, UK) 70–100

Jones R M, 1975, *Mechanics of Composite Materials* (Scripta Book Co, Washington, and McGraw-Hill)

Kedward K T (Editor), 1981, *Joining of Composite Materials: ASTM STP 749* (American Society for Testing and Materials, Philadelphia, USA)

Kendall D, 1996, Composites in communications in, *Proceedings 20th BPF Composites Congress,* Sept 1996 (British Plastics Federation, London) Session 3, paper 1

Kendall D, 1999, *The structural use of moulded composites: applications and opportunities,* Proceedings of a conference on Composites and Plastics in Construction, Nov 1999, BRE, Watford, UK (RAPRA Technology, Shawbury, Shrewsbury, UK) paper 19 1–10

Krenchel H, 1964, *Fibre Reinforcement* (Akademisk Forlag, Copenhagen, Denmark)

Kronenberg R, 1996, *Portable Architecture* (Architectural Press, Oxford) 30–35

Layton J, 1999, Weathering, in *Reinforced Plastics Durability* (Editor Pritchard G) (Woodhead Publishing, Abington UK) 186–218

Leggatt A J, 1984, *GRP and Buildings* (Butterworths, London)

Lubin G (Editor), 1982, *Handbook of Composites* (Van Nostrand Reinhold, New York)

Matthews F L and Rawlings R D, 1994, *Composite Materials: Engineering and Science* (Chapman and Hall, London)

Matthews F L, 1983, *Problems in the joining of GRP, Developments in GRP Technology* (Editor Harris B) (Elsevier Applied Science, London) 161–190

Matthews F L (Editor), 1987, *Joining Fibre-Reinforced Plastics* (Elsevier Applied Science, London)

Mayer R M (Editor), 1996, *Design of Composite Structures Against Fatigue: Applications to Wind Turbine Blades* (Mechanical Engineering Press, Bury St Edmonds, Suffolk)

Meier U and Winistörfer A, 1995, Retrofitting of structures through the external bonding of CFRP sheets, in *Proceedings of 2nd Intl RILEM Symp on Non-Metallic (FRP) Reinforcement for Concrete Structures FRPRCS-2,* Ghent, Belgium (Taerwe L, Editor) (E and FN Spon, London) 465–472

Melvin J, 1999, The Glass Menagerie, *Architects' Journal* 210 (3) 28–37

Moy S S J, Barnes F, Moriarty J, Dier A F, Kenchington A and Iverson B, 1999, *Structural up-grade and life extension of struts and beams using carbon-fibre-reinforced composites,* Proceedings of a conference on Composites and Plastics in Construction, Nov 1999, BRE, Watford, UK (RAPRA Technology, Shawbury, Shrewsbury) paper 15 1–10

Owen M J, 1970, *Fatigue, in Glass Reinforced Plastics* (Editor Parkyn B) (Iliffe Books, London) 251–267

Owen M J and Found M S, 1972, in *Solid/Solid Interfaces: Faraday Special Discussions no. 2* (Chemical Society, London) 77–89

Owen M J, 1974, Fatigue damage in glass-fibre-reinforced plastics, Chapter 7 in *Fracture and Fatigue* (Editor Broutman L J): Vol 5 of Composite Materials, Series (Editors Broutman L J and Krock R H) (Academic Press, New York and London) 313–369

Owen M J, 1974, Fatigue of carbon-fibre-reinforced plastics, Chapter 8 in *Fracture and Fatigue* (Editor Broutman L J): Volume 5 of Composite Materials, Series (Editors Broutman L J and Krock R H) (Academic Press, New York and London) 341–369

Pagano N J, (Editor), 1989, *Interlaminar Response of Composite Materials: Elsevier Composite Materials series, Volume 5*, (Elsevier Science Publishers BV, Amsterdam).

Parker D, 1995, Plastic surgery, *New Civil Engineer, issue 1128* (18 May) 21–23

Pawley M, 1993, *Future Systems: The Story of Tomorrow* (Phaidon Press, London)

Phillips D C and Harris B, 1977, *The strength, toughness and fatigue of composites, in Polymer Engineering Composites* (Editor MOW Richardson) (Applied Science Publishers, London) 45–154

Piggott M R and Harris B, 1980, Compression strength of carbon, glass and Kevlar 49 reinforced polyester resins, *J Mater Sci 15* 2523–2538

Powell P C, 1994, *Engineering with Fibre-Polymer Laminates* (Chapman and Hall, London)

Pritchard G (Editor), 1998, *Reinforced Plastics Durability* (Woodhead Publishing, Abington, UK)

Reifsnider K L (Editor), 1991, *Fatigue of Composite Materials: Composite Materials Series Volume 4* (Elsevier Amsterdam)

Rosner C N and Rizkalla S H, 1992, Design of bolted connections for orthotropic fibre-reinforced composite structural members, in *Advanced Composite Materials in Bridges and Structures* (Neale KW and Labossière, Editors) (Canadian Society for Civil Engineering, Montreal, Canada) 373–382

Schapery R A, 1968, Thermal expansion coefficients of composite materials based on energy principles, *J Compos Mater 2* 380–404

Scott Bader, 1994, *Crystic Polyester Handbook* (Scott Bader Company Ltd, Wollaston, Northants, UK)

Scott K A and Matthan J, 1970, Weathering, in *Glass Reinforced Plastics* (Editor Parkyn B), (Iliffe Books, London), 220–231

Stacey M, Voysey P, Potts A, Corden T and Garden H, 1999, *Development of a prototype modular railway canopy using advanced composites*, Proceedings of a conference on Composites and Plastics in Construction, Nov 1999, BRE, Watford, UK (RAPRA Technology, Shawbury, Shrewsbury, UK) paper 28 1–9

Summerscales J (Editor), 1987, *Non-Destructive Testing of Fibre-Reinforced Plastics Composites* (Elsevier Applied Science, London)

Talreja R, 1987, *Fatigue of Composite Materials* (Technomic Publishing Co Inc, Lancaster PA, USA)

Turvey G J, 2000, *Bolted connections in PFRP structures, Prog in Struct Engg Mater 2* 146–156

Weaver A, 1995, Bridge tests durability of rebar, *Reinforced Plastics 39 (7/8)* 20–22

Yang H H, 1993, *Kevlar Aramid Fibre* (John Wiley and Sons, Chichester)

8.2   KEY REFERENCE SOURCES OF FURTHER INFORMATION

Ashby M F and Jones D R H, 1980, *Engineering Materials, Vol 1* (Pergamon Press, Oxford)

Ashby M F, 1999, *Materials Selection in Mechanical Design (Second edition)* (Butterworth-Heinemann, Oxford)

Astrom B T, 1997, *Manufacturing of Polymer Composites* (Chapman and Hall, London)

Carlsson L A and Gillespie J W (Reviewing Editors), 1989, *Delaware Composites Design Encyclopaedia Volumes 1–6* (Technomic Publishing Co Inc, Lancaster, Pennsylvania, USA)

Clarke J L (Editor), 1996, *Structural Design of Polymer Composites: Eurocomp Design Code and Handbook* (E and FN Spon, Chapman and Hall London)

Eckold G C, 1994, *Design and Manufacture of Composite Structures* (Woodhead Publishing, Abington, UK)

Hancox N and Meyer R M, 1994, *Design Data for Reinforced Plastics* (Chapman and Hall, London)

Hollaway L C (Editor), 1990, *Polymers and Polymer Composites in Construction* (Thomas Telford Ltd, London)

Hollaway L C (Editor), 1994, *Handbook of Polymer Composites for Engineers* (Woodhead Publishing, Abington, UK)

Holmes M and Just D T, 1983, *GRP in Structural Engineering* (Applied Science Publishers, London)

Hull D and Clyne T W, 1996, *An Introduction to Composite Materials (second edition)* (Cambridge University Press, UK)

Hutchinson A R, 1997, *Joining of Fibre-Reinforced Polymer Composites*: CIRIA Project Report 46, (CIRIA, London)

Jones F R (Editor), 1994, *Handbook of Polymer-Fibre Composites* (Longman Green, Harlow, UK)

Kelly A (Editor), 1989, *Concise Encyclopaedia of Composite Materials* (Pergamon Press, Oxford)

Leggatt A J, 1984, *GRP and Buildings* (Butterworths, London)

Lubin G (Editor), 1982, *Handbook of Composites* (Van Nostrand Reinhold, New York)

8.3 OTHER GENERAL REFERENCES

Anon, 1998, *Composites for Infrastructure: A Guide for Civil Engineers* (Ray Publishing, USA)

Anon, no date, *Guidelines and Recommended Practice for Fibre-Glass Reinforced Plastic Architectural Products* (Composite Fabricators Association, USA)

Ball P, 1997, *Made to Measure: New Materials for the 21st Century* (Princeton University Press, Princeton, NJ, USA)

Carlsson L A and Gillespie J W (Reviewing Editors), 1989, *Composites Design Encyclopaedia Volumes 1–6*, (Technomic Publishing Co Inc, Lancaster, Pennsylvania, USA)

Crane F A A, Charles J A and Furness J, 1997, *Selection and Use of Engineering Materials* (Butterworth-Heinemann, Oxford)

Harris B, 1999, *Engineering Composite Materials (second edition)* (Institute of Materials, London)

Kelly A (Editor), 1989, *Concise Encyclopaedia of Composite Materials* (Pergamon Press, Oxford)

Lee S M, (Editor), 1984, *International Encyclopaedia of Composite Materials* (VCH Publishers New York)

Lubin G (Editor), 1982, *Handbook of Composites* (Van Nostrand Reinhold, New York)

Matthews F L and Rawlings R D, 1994, *Composite Materials: Engineering and Science* (Chapman and Hall, London)

Miller E, 1995, *Introduction to Plastics and Composites Mechanical Properties and Engineering Applications* (Marcel Dekker, New York)

Murphy J, 1998, *Reinforced Plastics Handbook (second edition)* (Elsevier Advanced Technology, Oxford)

Parkyn B (Editor), 1970, *Glass Reinforced Plastics* (Iliffe Books London)

Powell P C, 1994, *Engineering with Fibre-Polymer Laminates* (Chapman and Hall, London)

Richardson M O W (Editor), 1977, *Polymer Engineering Composites* (Applied Science Publishers, London)

Scott Bader, 1994, *Crystic Polyester Handbook* (Scott Bader Company Ltd, Wollaston, Northants, UK)

SP Systems, 1999, *Composite Materials Handbook* (SP Systems, Newport, IoW, UK)

Starr T, 1999, *Composites: A Profile of the Reinforced Plastics Industry-Markets and Suppliers to 2005* (Elsevier Advanced Technology, Oxford)

8.4   BIBLIOGRAPHY: OTHER REFERENCES TO SPECIFIC TOPICS

8.4.1   Composites applications

Ackermann G and Richter E, 1998, *Construction of chimneys and towers with polyester resin and glass fibre products*, Proceedings of 13th Congress of the International Association of Bridge and Structural Engineers, Helsinki, (IABSE, Zurich) 129–133

Allbones C, 1999, *The use of pultruded composites in the civil engineering and construction industry*, Proceedings of a conference on Composites and Plastics in Construction, Nov 1999, BRE, Watford, UK, (RAPRA Technology, Shawbury, Shrewsbury, UK) paper 4 1–10

Anon, 1993, Design and construction of the world's longest-span GRP bridge, *FRP International 1 (3)* 1–2

Anon, 1993, The longest fibreglass bridge in the USA, *FRP International, 1 (2)* 7

Anon, 1994, World's first composite lifting bridge, *FRP International 2 (2)* 5

Anon, 1998, FRC facades make formidable face-lifts, *Compos Technol 4* 15

Anon, 1998, Waste-water treatment: getting down and dirty with composites, *Compos Technol 4* 26–29

Anon, 1999, *Advanced Polymer Composites in Construction*: BRE information IP799 (Building Research Establishment, Watford, UK)

Anon, 1999, *Architectural Use of Polymer Composites*: BRE Digest 442, (Building Research Establishment, Watford, UK)

Aref A J and Parsons I D, 2000, Design and performance of a modular fibre reinforced plastic bridge, *Compos B: Engg 31* 619–628

Ballinger C, 1990, Structural FRP composites: Civil Engineering's material of the future? *ASCE Civ Engg 60 (7)* 63–65

Barno D S, 1997, *The future: architects should be aware of the benefits and solutions FRP (fibre-reinforced polymer) composites offer, Architectural Record, 187 (11)* 232

Bisanda E T N, 1993, The manufacture of roofing panels from sisal fibre-reinforced composites, *J Mater Processing and Technol 38* 369–379

Blachut J, 1998, Pressure vessel components: some recent developments in strength and buckling, *Prog in Struct Engg and Mater 1* 415–421

Brookes A J, 1998, *Building with composites, Architects' Journal 207 (24)* 37–38

Burford N K and Smith F W, 1999, *Bend it, shape it... creating an arched structure from a linear pultruded section*, Proceedings of a conference on Composites and Plastics in Construction, Nov 1999, BRE, Watford, UK, (RAPRA Technology, Shawbury, Shrewsbury, UK) paper 5 1–7

Buttrey D N, 1977, Applications of polymer composites in furniture and housewares, *Polymer Engineering Composites* (Editor MOW Richardson) (Applied Science Publishers, London) 493–533

Crowder J R, 1970, *Cladding and sheeting, in Glass Reinforced Plastics* (Editor Parkyn B) (Iliffe Books London) 81–93

Dodkins A R, 1993, *Design of displacement craft, Composite Materials in Maritime Structures – Vol. 2 : Practical Considerations* (Shenoi R A and Wellicome J F, Editors) (Cambridge University Press, UK) 3–25

Dodkins A R, Shenoi R A and Hawkins G L, 1994, Design of Joints and Attachments in *FRP Ships Structures, Marine Structures 7* 365–398

Dufton P W, 1999, *Lightweight Thermoset Composites* (RAPRA Technology Ltd, Shawbury, UK)

Ehlen M A 1999, Life-cycle costs of fibre-reinforced-polymer bridge decks, *J Mater in Civ Engg 11* 224–230

El Mikawi M and Mosallam A S, 1996, A methodology for evaluation of the use of advanced composites in Structural civil engineering applications, *Compos B: Engg 27B* 203–215

Farmer N and Gee A, 1998, *FRPs – New Millennium, New Materials, Concrete Engg Internat 2* 29–31

Foster D C, Richards D and Boigner B R, 2000, Design and installation of a fibre-reinforced polymer composite bridge, *J Compos for Construction 4* 33–37

Fujimori T, Sugizaki K 1998, Large-span structural system using new materials, *J Mater in Civ Engg 10* 203–207

Ganga Rao H V S, Thippeswamy H K, Shekar V and Craigo C 1999, Development of class fibre-reinforced polymer composite bridge deck, *SAMPE Journal 35* 12–24

Gibson A G, 1993, *Composites in offshore structures, Composite Materials in Maritime Structures – Vol. 2 : Practical Considerations* (Shenoi R A and Wellicome J F, Editors) (Cambridge University Press, UK) 199–228

Goldsworthy W B and Davidson M, 1994, Composites pioneer lends perspective for shaping the industry's future, *Advanced Materials and Processes 146* 70–71

Goldsworthy W B, Rutan B and Hiel C, 1997, From complete chaos to clear concepts –3, *SAMPE Journal 33* 17–25

Harvey W J, 1993, A reinforced-plastic footbridge, Aberfeldy, UK, *Struct Engg Intl 4* 229–232

Head P R, 1988, *Use of fibre-reinforced plastics in bridge structures*, Proceedings of 13th Congress of the International Association of Bridge and Structural Engineers, Helsinki (IABSE, Zurich) 123–128

Head P R, 1997, Civil composites, *Materials World 5* 83–85

Holmes M and Just D T, 1983, *GRP in Structural Engineering* (Applied Science Publishers, London)

Ikeda K, Sekijima K and Okamura H, 1988, *New materials for tunnel supports*, Proceedings of 13th Congress of the International Association of Bridge and Structural Engineers, Helsinki (IABSE, Zurich) 27–32

Iyer S L, 1993, *Advanced composite demonstration bridge deck, Fibre-Reinforced Plastic Reinforcement for Reinforced Concrete Structures* (FRPRCS–1) SP 138 (Editors Nanni A and Dolan C W) (American Concrete Institute, Michigan, USA) 831

Karbhari V M, Seible F, Burgueño R, Davol A, Wernli M and Zhao L, 1998, *Structural characterization of fibre-reinforced composite short- and medium-span bridge systems*, Proceedings of the 8th European Conference on Composite Materials ECCM8, Naples, June 1998, Symposium 3, Volume 2 (Woodhead Publishing, Abington, UK) 35–42

Karbhari V M, Seible F, Burgueño R, Davol A, Wernli M and Zhao L, 2000, Structural characterization of fibre-reinforced composite short- and medium-span bridge systems, *Appl Compos Mater 7* 151–182

Keller T, 2000, *Bridges and buildings in the context of materialisation*, Proceedings of 16th Congress of the International Association of Bridge and Structural Engineers, Lucerne, Switzerland (IABSE, Zurich) paper 109 8–9

Khalifa M A, Hodhod O A and Zaki M A 1996, Analysis and design methodology for an FRP cable-stayed pedestrian bridge, *Compos B: Engg 27* 307–317

Kim D H, 1995, *Composites Structures for Civil and Architectural Engineering* (E and FN Spon, London)

Krikanov A A, 2000, Composite pressure vessels with higher stiffness, *Compos Structures 48* 119–127

Leggatt A J, 1976, G R P and buildings, *The Structural Engineer 54 (12)* 479–486

Meier U, 1987, Proposal for a carbon-fibre-reinforced composite bridge across the Strait of Gibraltar at its narrowest site, *Prod Inst Mechanical Engineering, 201*, paper B2.

Miller E, 1995, *Introduction to Plastics and Composites Mechanical Properties and Engineering Applications* (Marcel Dekker, New York)

Mitchell R G B, 1970, in *Glass Reinforced Plastics* (Editor Parkyn B) (Iliffe Books London) 11–29

Mosallam A S, Dutta P K and Hui D, (Editors), 1996, *Compos B: Engg, 27B (3/4) Structural Composites in Infrastructures* (Journal Special Issue) 203–385

Mosallam A S and Hui D (Editors), 2000, *Compos B: Engg, 31B* (6/7) 443–444 *Infrastructure Composites* (Journal Special Issue) 443–628

Neale K W and Labossière P (Editors), 1992, *Advanced Composite Materials in Bridges and Structures* (Canadian Society for Civil Engineering, Montreal, Canada)

Noisternig J F and Jungwirth D, 1998, *GRP and CFRP elements for the construction industry,* Proceedings of the 8th European Conference on Composite Materials ECCM8, Naples, June 1998, Symposium 3, Volume 2 (Woodhead Publishing, Abington, UK) 293–300

Noisternig J F, 2000, Carbon fibre composites as stay cables for bridges, *Appl Compos Mater 7* 139–150

Norwood L S, 1999, *Glass-reinforced plastic for construction*, Proceedings of a conference on Composites and Plastics in Construction, Nov 1999, BRE, Watford, UK (RAPRA Technology, Shawbury, Shrewsbury, UK) paper 3 1–11

Pascoe M W, 1977, Applications of polymer composites in building and construction, *Polymer Engineering Composites* (Editor MOW Richardson) (Applied Science Publishers, London) 419–439

Piccioli E, 1998, *Composites in construction*, Proceedings of the 8th European Conference on Composite Materials ECCM8, Naples, June 1998, Symposium 3, Volume 2, (Woodhead Publishing, Abington, UK) 27–34

Rawson J, 1994, Fibre-reinforced plastics components, *Architects' Journal 200 (13)* 43–45

Saadatmanesh H, 1994, Fibre composites for new and existing structures, *ACI Structural Journal 91* 346–354

Saadatmanesh H and Ehsani M R (Editors), 1998, *Fibre Composites in Infra-structure (Volume 1)* (University of Arizona Press, Tucson, Arizona)

Schurter U, 2000, *Storchenbrücke Winterthur-a cable-stayed bridge* Proceedings of 16th Congress of the International Association of Bridge and Structural Engineers, Lucerne, Switzerland (IABSE, Zurich) paper 240 126–127

Scott D V, 1996, *FRP Pipes and Structures: A Survey of European Markets and Suppliers* (Elsevier Advanced Technology, Oxford)

Scott D V, 2000 (Elsevier Advanced Technology, Oxford)

Sheard P, Clarke J L and Pilakoutas K, 1999, *The application of non-ferrous reinforcement in civil engineering*, Proceedings of a conference on Composites and Plastics in Construction, Nov 1999, BRE, Watford, UK (RAPRA Technology, Shawbury, Shrewsbury, UK) paper 11 1–7

Shenoi R A and Wellicome J F (Editors), 1993, *Composite Materials in Maritime Structures – Vol 1: Fundamental Aspects* (Cambridge University Press, UK)

Shenoi R A and Wellicome J F (Editors), 1993, *Composite Materials in Maritime Structures – Vol 2: Practical Considerations* (Shenoi R A and Wellicome J F, Editors) (Cambridge University Press, UK)

Shenoi R A and Moy S S J, 1999, *Engineering challenges in the use of advanced composites in made-to-order products* Proceedings of a conference on Composites and Plastics in Construction, Nov 1999, BRE, Watford, UK, (RAPRA Technology, Shawbury, Shrewsbury, UK) paper 4 1–6

Smith S J, Bank L C, Gentry T R, Nuss K H, Hurd S H and Duich S J, 2000, Analysis and testing of a prototype pultruded composite causeway structure, *Compos Structures 49* 141–150

Srinivasa L I and Rajan S (Editors), 1991, *Advanced Composites Materials in Civil Engineering Structures,* Proceedings of a Speciality Conference, Las Vegas (American Society of Civil Engineers, USA)

Starnini S, Arduini M, Gelsomino L and Nanni A, 1998, *FRP composites for modular construction of housing projects*, Proceedings of the 8th European Conference on Composite Materials ECCM8, Naples, June 1998, Symposium 3, Volume 2, (Woodhead Publishing, Abington, UK) 78–82

Starr T, 1998, Get the strength of GRP around you, *Reinforced Plastics June* 42–46

Tabiei A, Svenson A, Hargarve M and Bank L 1998, Impact performance of pultruded beams for highway safety applications, *Compos Structures 42* 231–237

Taerwe L and Clarke J L (Editors), 1996, *Proceedings of conference on Advanced Composite Materials in Bridges and Structures, Montreal Canada* (Canadian Society for Civil Engineering, Montreal, Canada)

Thornburrow P, 1992, *Bridge over the River Tay, Vetrotex GRP Review*, (Vetrotex, Wallingford, Oxon, UK)

Tsuji Y, Kanda M and Tamura T, 1993, Applications of FRP materials to prestressed concrete bridges and other structures in Japan, *PCI Journal, Jul/Aug 50*

Van Erp G M, 1999, A new fibre composite beam for civil engineering applications *Adv Compos Let 8* 219–225

Wallace S, 1999, *Design of falls creek trail bridge – A fibre-reinforced-polymer composite bridge*, Transportation Research Record, no 1652, 133–142

Weaver A, 1997, Kolding bridge: a technical landmark, *Reinforced Plastics, 41 (8)* 30–33

Winter J, 1993, New use for composite materials, *Architects' Journal 197 (5)* 39–40

8.4.2   Codes of practice

Anon, 1999, Review of Standards and Codes for Fibreglass Composites, *Compos Technol 5* 16–20

BSI, 1987, *Specification for Design and Construction of Vessels and Tanks in Reinforced Plastics*: BS 4994 (British Standards Institute, London)

BSI, 1999, *Reinforced Plastics Composites – Specification for Pultruded Profiles: parts 1,2 and 3* (British Standards Institute, London)

Clarke J L, (Editor), 1996, *Structural Design of Polymer Composites: Eurocomp Design Code and Handbook* (E and FN Spon, Chapman and Hall London)

Curtis P T (Editor), 1988, *CRAG Test Methods for the Measurement of Engineering Properties of Fibre-Reinforced Plastics Composites* (Tech Report 88012) (Ministry of Defence Procurement Executive, Farnborough, UK)

Sims G D, 1999, *European standardisation: composite materials and pultruded profiles,* Proceedings of a conference on Composites and Plastics in Construction, Nov 1999, BRE, Watford, UK (RAPRA Technology, Shawbury, Shrewsbury, UK) paper 7 1–8

8.4.3   Concrete-related applications (rebars, strengthening etc)

Al Zahrani M M, Al Dulaijan S U, Nanni A, Bakis C E and Boothby T E 1999, Evaluation of bond using FRP rods with axisymmetric deformations, *Construction and Building Mater 13* 299–309

Alsayed S H, Al-Salloum Y A and Almusallam T H, 2000, Performance of glass-fibre-reinforced-plastic bars as a reinforcing material for concrete structures, *Compos B: Engg 31* 555–567

Alsayed S H and Alhozaimy A M 1999, Ductility of concrete beams reinforced with FRP bars and steel fibres, *J Compos Mater 33* 1792–1806

American Concrete Institute, 1996, *Fibre-Reinforced-Plastic Reinforcement for Concrete Structures: ACI Committee 544 State-of-the-art Report* (American Concrete Institute, Michigan, USA)

Andersen J, 1994, *Fibre-reinforced composite materials for reinforced concrete construction*, Proceedings of Intl. Seminar on Building the Future: Innovation in Design Materials and Construction, Brighton April 1993 (Editors Gara F K, Armer G S T and Clarke J L) (E and FN Spon, London) 211–220

Anon, 1993, Carbon-fibre strands pre-stress Calgary span, *Engineering News Record Oct 1993 21*

Anon, 1994, *Design and Construction Guidelines for Pre-stressed Concrete Highway Bridges using FRP Tendons* (Sumitomo and Mitsui Construction Companies, Japan)

ARC Slimline, 1979, *The Specification and Performance of Slimline Glass-Fibre-Reinforced Concrete Pipes* (ARC Slimline)

Arockiasamy M, Chidambaram S, Amer A and Shahawy M A, 2000, Time-dependent deformations of concrete beams reinforced with CFRP bars, *Compos B: Engg 31*, 577–592

Bakis C E, Uppuluri V S, Nanni A and Boothby T, 1998, Analysis of bonding mechanisms of smooth and lugged FRP rods embedded in concrete, *Compos Sci and Technol 58* 1307–1319

Benedetti A and Nanni A, 1998, *On carbon-fibre strengthening of heat-damaged pre-stressed concrete*, Proceedings of the 8th European Conference on Composite Materials ECCM8, Naples, June 1998, Symposium 3, Volume 2 (Woodhead Publishing, Abington, UK) 67–74

Benmokrane B, Chaallal O and Masmoudi R, 1995, Glass-fibre-reinforced plastic (GRP) rebars for concrete structures, *Construction and Building Mater 9* 353–364

Benmokrane B, Tighiouart B and Chaallal O, 1996, Bond strength and load distribution of composite GRP reinforcing bars in concrete, *ACI Materials Journal 93* 246–253

Benmokrane B, Zhang B R and Chennouf A, 2000, Tensile properties and pullout behaviour of AFRP and CFRP rods for grouted anchor applications, *Construction and Building Mater 14* 157–170

Boyle H C and Karbhari V M 1995, Bond and behaviour of composite reinforcing bars in concrete, *Polymer-Plastics Technol and Engg 34* 697–720

Burgoyne C J, 1993, *Should FRP be bonded to concrete? Fibre-Reinforced Plastic Reinforcement for Concrete Structures* (FRPRCS-1) SP 138 (Editors Nanni A and Dolan C W) (American Concrete Institute, Michigan, USA) 367

Burgoyne C J, 1997, *Rational use of advanced composites in concrete*, Proceedings of 3rd Intl RILEM Symposium on Non-Metallic (FRP) Reinforcement for Concrete Structures FRPRCS-3, Sapporo, Japan (Taerwe L, Editor) (Japan Concrete Society, Tokyo, Japan) 75–88

Canning L, Hollaway L C and Thorne A M, 1999, An investigation of the composite action of an FRP/concrete prismatic beam, *Construction and Building Mater 13* 417–426

Casas J R and Aparicio A C, 1990, *A full-scale experiment on a prestressed concrete structure with high-strength fibres; the North ring road in Barcelona*, Proceedings of the XI International Congress of FIP, Hamburg, June 1990 (Fédération Internationale de la Précontrainte, London) Vol 2 T15–T18

Chaallal O and Benmokrane B, 1996, Fibre-reinforced plastic rebars for concrete applications, *Compos B: Engg 27B* 245–252

Clarke J L, 1993, *Alternative Materials for the Reinforcement and Pre-Stressing of Concrete* (Blackie Academic and Professional, Glasgow, UK)

Clarke J L, 1994, *Fibre composites for the reinforcement of concrete*, Proceedings of an Intl. Seminar on Building the Future: Innovation in Design Materials and Construction Brighton April 1993 (Editors Gara F K, Armer GST and Clarke J L) (E and FN Spon, London) 183–191

Clarke J L and Waldron P, 1996, The reinforcement of concrete structures with advanced composites, *The Structural Engineer 74* 283

Clarke J L, 1998, Steeling concrete with fibre-reinforced plastics, *Materials World 6* 78–80

Clarke J L, 1999, *FRP materials for reinforcing and strengthening concrete structures*, Proceedings of a conference on Composites and Plastics in Construction, Nov 1999, BRE, Watford, UK (RAPRA Technology, Shawbury, Shrewsbury, UK) paper 17 1–6

Czaderski C, 2000, *Shear strengthening of reinforced concrete with CFRP* Proceedings of 16th Congress of the International Association of Bridge and Structural Engineers, Lucerne, Switzerland (IABSE, Zurich) paper 164 (376–377)

Dolan C W, Rizkalla S H and Nanni A (Editors), 1999, *Fibre-Reinforced Polymer Reinforcement for Reinforced Concrete Structures* (FRPRCS–4) SP188 (American Concrete Institute, Michigan, USA) pp 1464

El Ghandour A W, Pilakoutas K and Waldron P, 1998, *Behaviour of FRP RC slabs with CFRP shear reinforcement*, Proceedings of the 8th European Conference on Composite Materials ECCM8, Naples, June 1998, Symposium 3, Volume 2 (Woodhead Publishing, Abington, UK) 399–406

Franke L and Wolfe R, 1988, *Fibre-glass tendons for pre-stressed concrete bridges*, Proceedings of 13th Congress of the International Association of Bridge and Structural Engineers, Helsinki (IABSE, Zurich) 51–56

Garden H N and Hollaway L C, 1998, An experimental study of the influence of plate end anchorage of carbon fibre composite plates used to strengthen reinforced concrete beams, *Compos Structures 42* 175–188

Gendron G, Picard A and Guerin M C 1999, A theoretical study on shear strengthening of reinforced concrete beams using composite plates, *Compos Structures 45* 303–309

Geng Z J, Chajes M J, Chou T W and Pan D Y C 1998, The retrofitting of reinforced concrete column-to-beam connections, *Compos Sci and Technol 58* 1297–1305

Grace N F, Sayed G A, Soliman A K and Saleh K R, 1999, Strengthening reinforced concrete beams using fibre reinforced polymer (FRP) laminates, *ACI Structural Journal 96* 865–874

Grace N F, Sayed G A, Wahba J and Sakla S, 1999, Mathematical solution for carbon fibre-reinforced polymer prestressed concrete skew bridges, *ACI Structural Journal 96* 981–987

Greco C, Manfredi G, Pecce M and Realfonzo R, 1998, *Experimental analysis of bond between deformed GRP rebars and concrete*, Proceedings of the 8th European Conference on Composite Materials ECCM8, Naples, June 1998, Symposium 3, Volume 2, (Woodhead Publishing, Abington, UK) 301–308

Guo J and Cox J V, 2000, An interface model for the mechanical interaction between FRP bars and concrete, *J Reinf Plastics and Compos 19* 15–33

Hollaway L C, Canning L and Thorne A M, 1999, *An investigation into the load characteristics of a unique advanced polymer-composite/concrete beam*, Proceedings of a conference on Composites and Plastics in Construction, Nov 1999, BRE, Watford, UK (RAPRA Technology, Shawbury, Shrewsbury, UK) paper 10 1–8

Hollaway L C and Leeming M (Editors), 1999, *Strengthening of Reinforced Concrete Structures* (Woodhead Publishing, Abington, UK)

I Struct E, 1999, *Interim guidance on the Design of Reinforced-Concrete Structures using Fibre-Composite Reinforcement* (Institute of Structural Engineers Press, London)

Ibell T J and Burgoyne C J, 1999, Use of fibre-reinforced plastics versus steel for shear reinforcement of concrete, *ACI Structural Journal 96* 997–1002

Jensen D W and Meade J G, 1999, Flexure of concrete with chopped GR/EP prepreg fibres, *J Reinf Plastics and Compos 18* 1496–1515

JSCE, 1996, *State-of-the-art report on continuous fibre reinforcing materials*, (Japanese Society of Civil Engineers, Tokyo)

JSCE, 1996, *Recommendation for design and construction of concrete structures using continuous fibre reinforcing materials* (Japanese Society of Civil Engineers, Tokyo)

Kachlakev DI, 2000, Experimental and analytical study on unidirectional and off-axis GRP rebars in concrete, *Compos B: Engg 31* 569–575

Konsta-Gdoutos M and Karayiannis C 1998, Flexural behaviour of concrete beams reinforced with FRP bars, *Adv Compos Letters 7* 133–137

Lin A, Guo X C and Lu Z T, 2000, *Experimental study of beams with CFRP strengthening*, Proceedings of 16th Congress of the International Association of Bridge and Structural Engineers, Lucerne, Switzerland (IABSE, Zurich) paper 43 (366–367)

Mahmoud Z I, Rizkalla S H and Zaghloul E E R, 1999, Transfer and development lengths of carbon-fibre-reinforced polymers pre-stressing reinforcement, *ACI Structural Journal 96* 594–602

Mirmiran A and Shahawy M A, 1996, A new concrete-filled hollow FRP composite column, *Compos B: Engg, 27B* 263–268

Morita S (Editor), 1997, Proceedings of the 3rd RILEM International conference on *Non-Metallic (FRP) Reinforcement for Concrete Structures* (FRPRCS–3) Sapporo, Japan, Oct 1997 (Japan Concrete Society, Tokyo, Japan)

Morris W, Vazquez M and De Sanchez S R, 2000, Efficiency of coatings applied on rebars in concrete, *J Mater Sci 35* 1885–1890

Nanni A and Dolan C W (Editors), 1993, *Fibre-Reinforced Plastic Reinforcement for Concrete Structures* (FRPRCS–1) SP 138 (American Concrete Institute, Michigan, USA)

Noritake K *et al*, 1993, Practical applications of aramid FRP rods to pre-stressed concrete structures, in *Fibre-Reinforced Plastic Reinforcement for Concrete Structures* (FRPRCS–1) SP 138 (Editors Nanni A and Dolan C W) (American Concrete Institute, Michigan, USA) 853

Park S Y and Naaman A E, 1999, Dowel behaviour of tensioned fibre-reinforced polymer (FRP) tendons, *ACI Structural Journal 96* 799–806

Parvin A and Granata P, 2000, Investigation on the effects of fibre composites at concrete joints, *Compos B: Engg, 31* 499–509

Pecce M, Manfredi G and Cosenza E, 1998, *Experimental behaviour of concrete beams reinforced with GRP bars*, Proceedings of the 8th European Conference on Composite Materials ECCM8, Naples, June 1998, Symposium 3, Volume 2 (Woodhead Publishing, Abington, UK) 227–234

Ramana V P V, Kant T, Morton S E, Dutta P K, Mukherjee A and Desai Y M, 2000, Behaviour of CFRPC strengthened reinforced concrete beams with varying degrees of strengthening, *Compos B: Engg 31* 461–470

Saafi M and Toutanji H, 1998, Flexural capacity of prestressed concrete beams reinforced with aramid fibre-reinforced polymer (AFRP) rectangular tendons, *Construction and Building Mater 12* 245– 249

Shahawy M A, Beitelman T, Arockiasamy M and Sandepudi K S, 1996, Static flexural response of members pre-tensioned with multiple layered aramid fibre tendons, *Compos B: Engg 27B* 253–261

Shahawy M A, Mirmiran A and Beitelman T, 2000, Tests and modelling of carbon-wrapped concrete columns, *Compos B: Engg 31* 471–480

Soudki K A, 1998, FRP reinforcement for pre-stressed concrete structures, *Prog in Struct Engg and Mater 1* 135–142

Stoll F, Saliba J E, Casper L E, 2000, Experimental study of CFRP-prestressed high-strength concrete bridge beams, *Compos Structures 49* 191–200

Sumitani A, Kikuchi M Sotooka M, Akimoto H and Ozawa N 1998, Pultruded CFRP rods for ground anchor application, *Adv Compos Mater 7* 395–401

Taerwe L (Editor), 1995, *Proceedings of 2nd Intl RILEM Symp on Non-Metallic (FRP) Reinforcement for Concrete Structures* FRPRCS–2, Ghent, Belgium (E and FN Spon, London)

Tanigaki M, Okamoto T, Tamura T, Matsubara S and Nomura S, 1988, *Study of braided aramid-fibre rods for reinforcing concrete*, Proceedings of 13th Congress of the International Association of Bridge and Structural Engineers, Helsinki (IABSE, Zurich) 15–20

Tighiouart B, Benmokrane B and Gao D, 1998, Investigation of bond in concrete member with fibre-reinforced polymer (FRP) bars, *Construction and Building Mater 12* 453–462

Tighiouart B, Benmokrane Band Mukhopadhyaya P, 1999, Bond strength of glass FRP rebar splices in beams under static loading, *Construction and Building Mate, 13* 383–392

Uomoto T, 1995, Concrete composites in the construction field, *Adv Compos Mater 4* 261–269

Wolff R and Meisseler H J, 1993, Glass-fibre pre-stressing system, *Alternative Materials for the Reinforcement and Pre-Stressing of Concrete* (J L Clarke, Editor) (Blackie Academic and Professional, Glasgow, UK) 127

Zenon A and Kypros P, Peter W, 1998, *Bond of FRP anchorages and splices in concrete beams*, Proceedings of the 8th European Conference on Composite Materials ECCM8, Naples, June 1998, Symposium 3, Volume 2 (Woodhead Publishing, Abington, UK) 277–284

Zoch P *et al*, 1991, *Carbon-fibre composite cables: a new class of pre-stressing members*, Proceedings of the 70th Annual Convention of the Transportation Research Board Washington DC Jan 1991, (Transportation Research Board of NRC, Washington).

8.4.4   Design and analysis

Bouverie R P, 1970, Design logic for GRP in buildings, in *Glass Reinforced Plastics* (Editor Parkyn B) (Iliffe Books, London) 285–295

Davalos J F, Salim H A, Qiao P, Lopez A and Barbero E J 1996, Analysis and design of pultruded FRP shapes under bending, *Compos B: Engg 27B* 295–305

de Wilde W P and Blain W R (Editors), 1990, *Composite Materials: Design and Analysis* (Editors de Wilde W P and Blain W R) (Springer-Verlag, Berlin)

Gan L H, Ye L and Mai Y W 1999, Optimum design of cross-sectional profiles of pultruded box beams with high ultimate strength, *Compos Structures 45* 279–288

Gan L H, Ye L and Mai Y W, 1999, Design and evaluation of various section profiles for pultruded deck panels, *Compos Structures 47* 719–725

Hardy S J and Malik N H, 1990, *Optimum design of composite-reinforced pressure vessels*, Proceedings of an Intl Conf on Applied Stress Analysis (Hyde T H and Ollerton E, Editors) Nottingham, Aug 30–31 (Elsevier Applied Science, London) 429–438

Kaelble D H, 1985, *Computer-Aided design of Polymers and Composites* (Marcel Dekker, New York)

Martin P M J W, 1990, Filament-wound composite pressure vessels: computer-aided optimisation and manufacturing, in *Composite Materials: Design and Analysis* (Editors de Wilde W P and Blain W R) (Springer-Verlag, Berlin) 465–473

Mottram J T and Zheng Y, 1996, State-of-art review of the design of beam-to-column connections for pultruded frames, *Compos Structures, 35* 387–401

Qiao P, Davalos J F and Brown B, 2000, A systematic analysis and design approach for single-span FRP deck/stringer bridges, *Compos B: Engg 31* 593–609

Quinn J A, 1995, *Composites Design Manual*, (James Quinn Associates, Liverpool, UK)

Richardson T, 1987, *Composites: A Design Guide* (Industrial Press Inc, New York)

Soden P D and Eckold G C, 1983, *Design of GRP pressure vessels, Developments in GRP Technology* (Editor Harris B) (Applied Science Publishers, London) 91–159

Surrel Y, Vautrin A and Verchery G (Editors), 1990, *Analysis and Design of Composite Materials Structures* (Editions Pluralis, Paris)

Tetlow R, 1973, Structural engineering design and applications, in *Carbon Fibres in Engineering* (Editor Langley M) (McGraw Hill (UK) Ltd Maidenhead, UK) 108–159

8.4.5 Durability and damage

Bogner B, 1999, Survey of long-term durability of fibreglass-reinforced plastics tanks and pipes, in *Reinforced Plastics Durability* (Editor Pritchard G) (Woodhead Publishing, Abington UK) 267–281

Cardon A, (Editor), 1996, *Durability Analysis of Structural Composite Systems* (AA Balkema, Rotterdam)

Chin J W, Nguyen T and Aouadi K 1997, Effects of environmental exposure on fibre-reinforced plastic (FRP) materials used in construction, *J Compos Technol and Research 19* 205–213

Coomarasamy A and Goodman S, 1999, Investigation of the durability characteristics of fibre-reinforced-plastic (FRP) materials in a concrete environment, *J Thermoplastic Compos Mater 12* 214–226

Crowder J R and Howard C M, 1990, End-use performance and time-dependent characteristics, *Polymers and Polymer Composites in Construction* (Editor Hollaway L C) (Thomas Telford Ltd, London) 139–165

Dawson D, 2000, FRP tanks fight corrosion and contamination, *Compos Technol 6* 20–24

Dutta P K and Hui D 1996, Low-temperature and freeze-thaw durability of thick composites, *Compos B: Engg 27B* 371–379

Emri I, 1996, Time-dependent phenomena related to the durability analysis of composite structures, in *Durability Analysis of Structural Composite Systems* (AA Balkema, Rotterdam) 85–122

Hamilton H R and Dolan C W, 2000, Durability of FRP reinforcements for concrete, *Prog in Struct Engg and Mater 2* 139–145

Highsmith A L and Reifsnider K L, 1982, Stiffness-reduction mechanisms in composite laminates, in *Damage in Composite Materials*: ASTM STP 775 (Editor Reifsnider K L) (American Society for Testing and Materials Philadelphia USA) 103–117

Karbhari V M and Eckel D A, 1994, Effect of cold regions climate on composite jacketed concrete columns, *J Cold Regions Engg 8* 73–86

Karbhari V M and Eckel D A, 1995, Effects of short-term environmental exposure on axial strengthening capacity of composite jacketed concrete, *J Compos Technol and Research 17* 99–106

Liggatt J J, Pritchard G and Pethrick R A, 1999, Temperature–its effects on the durability of reinforced plastics, in *Reinforced Plastics Durability* (Editor Pritchard G) (Woodhead Publishing Abington UK) 111–150

Masters J E and Reifsnider K L, 1982, An investigation of cumulative damage development in quasi-isotropic graphite/epoxy laminates, in *Damage in Composite Materials* (ASTM STP 775) (Editor Reifsnider K L) (American Society for Testing and Materials Philadelphia USA)

Mukhopadhyaya P, Swamy R N and Lynsdale C J, 1998, Influence of aggressive exposure conditions on the behaviour of adhesive bonded concrete-GRP joints, *Construction and Building Mater 12* 427–446

Phoenix SL 2000, Modelling the statistical lifetime of glass-fibre/polymer matrix composites in tension, *Compos Structures 48* 19–29

Read P J C L and Shenoi R A, 1995, A review of fatigue damage modelling in the context of marine FRP laminates, *Marine Structures 8* 257–278

Reifsnider K L (Editor), 1982, *Damage in Composite Materials* (ASTM STP 775) (American Society for Testing and Materials Philadelphia USA)

Saccani A and Magnaghi V, 1999, Durability of epoxy resin-based materials for the repair of damaged cementitious composites, *Cement and Concrete Research 29* 95–98

Sen R, Shahawy M A and Sukumar S, Rosas J, 1999, Durability of carbon-fibre-reinforced polymer (CFRP) pre-tensioned elements under tidal/thermal cycles, *ACI Structural Journal 96* 450–457

Sen R, Shahawy M A, Mullins G and Spain J, 1999, Durability of carbon-fibre-reinforced polymer/epoxy/concrete bond in marine environment, *ACI Structural Journal 96* 906–914

Smith W S, 1979, *Environmental Effects on Aramid Composites*, Proceedings of the December 1979 SPE Meeting, Los Angeles (Society for Plastics Engineering) paper A-34

Talreja R (Editor), 1994, *Damage Mechanics of Composite Materials: Composite Materials Series Vol 9* (Elsevier Science BV, Amsterdam)

Tannous F E and Saadatmanesh H 1998, Environmental effects on the mechanical properties of E-glass FRP rebars, *ACI Materials Journal 95* 87–100

Toutanji H A and Goodman S, 1999, Durability characteristics of concrete columns confined with advanced composite materials, *Compos Structures 44* 155–161

Zhang S, Ye L and Mai Y W 1999, Effects of saline water immersion on glass-fibre/vinyl-ester wrapped concrete columns, *J Reinf Plastics and Compos 18* 1592–1604

Zoghi M, Casper L and Müller T, 1998, *The effect of UV light on advanced composite material for the Tech21 bridge*, Proceedings of the 8th European Conference on Composite Materials ECCM8, Naples, June 1998, Symposium 3, Volume 2 (Woodhead Publishing, Abington, UK) 397–398

8.4.6   Fire properties and performance

Mouritz A P and Mathys Z, 1999, Post-fire mechanical properties of marine polymer composites, *Compos Structures 47* 643–653

Tyberg C S, Sankarapandian M, Bears K and Shih P, Loos AC, Dillard D and McGrath J E, 1999, Tough, void-free, flame retardant phenolic matrix materials, *Construction and Building Mater 13* 343–353

Williams J D, 1977, Aspects of the fire properties of polymer composites, in *Polymer Engineering Composites* (Editor MOW Richardson) (Applied Science Publishers, London) 363–418

8.4.7 Joining

Abdel Naby S F M and Hollaway L C 1993, The experimental behaviour of bolted joints in pultruded glass/polyester material: I Single-bolt joints, *Composites 24* 531–538

Abdel Naby S F M and Hollaway L C 1993, The experimental behaviour of bolted joints in pultruded glass/polyester material: II Two-bolt joints *Composites 24* 539–546

Adams R D and Wake W C, 1994, *Structural Adhesive Joints in Engineering* (Elsevier Applied Science, London)

Cooper C and Turvey G J, 1995, Effects of joint geometry and bolt torque on the structural performance of single-bolt tension joints in pultruded GRP sheet material, *Compos Structures 32* 217–226

Erki M A, 1995, Bolted glass-fibre-reinforced plastic joints, *Canadian J Civ Engg 22* 736–744

Goldsworthy W B and Hiel C, 1998, Composite structures are a snap, *SAMPE Journal 34* 24–30

Grutta J T, Miskioglu I, Charoenphan S and Vable M, 2000, Strength of bolted joints in composites under concentrated moment, *J Compos Mater 34* 1242–1262

Hart-Smith L J, 1987, Joints in composites, in *Engineered Materials Handbook Volume 1* (ASM International, Ohio, USA) 479–495

Hassan N K, Mohamedien M A and Rizkalla S H 1996, Finite element analysis of bolted connections for PFRP composites, *Compos B: Engg 27* 339–349

Her S C, 1999, Stress analysis of adhesively-bonded lap joints, *Compos Structures 47* 673–678

Hollaway L C, 1990, Adhesive and bolted joints, in *Polymers and Polymer Composites in Construction* (Editor Hollaway L C), (Thomas Telford Ltd, London), Chapter 6 107–137

Hutchinson A R, 1999, *Adhesively bonded joints involving fibre-reinforced polymer composites*, Proceedings of a conference on Composites and Plastics in Construction, Nov 1999, BRE, Watford, UK (RAPRA Technology, Shawbury, Shrewsbury, UK) paper 9 1–10

Lie S T, Yu G and Zhao Z, 2000, Analysis of mechanically fastened composite joints by boundary element methods, *Compos B: Engg, 31* 693–705

Mosallam A S, Abdel Hamid M K and Conway J H, 1994, Performance of pultruded FRP connectors under static and dynamic loads, *J Reinf Plastics and Compos 13* 386–407

Parker B M, 1983, The effect of composite pre-bond moisture on adhesive bonded joints, *Composites 14* 226

Prabhakaran R, Razzaq Z and Devara S 1996, Load and resistance factor design (LRFD) approach for bolted joints in pultruded composites, *Compos B: Engg 27B* 351–360

Rosner C N and Rizkalla S H, 1995, Bolted connections for fibre-reinforced composite structural members – experimental program, *J Mater in Civ Engg 7* 223–231

Shenoi R A and Hawkins G L, 1993, Practical design of joints and attachments, *Composite Materials in Maritime Structures – Vol. 2 : Practical Considerations* (Shenoi R A and Wellicome J F, Editors) (Cambridge University Press, UK) 63–90

Smith S J, Parsons I D and Hjelmstad K D 1998, An experimental study of the behaviour of connections for pultruded GRP I-beams and rectangular tubes, *Compos Structures 42* 281–290

Turvey G J, 1998, Single-bolt tension joint tests on pultruded GRP plate – effects of tension direction relative to pultrusion direction, *Compos Structures 42* 341–351

Winkle I E, 1993, The rôle of adhesives, *Composite Materials in Maritime Structures – Vol. 2 : Practical Considerations* (Shenoi R A and Wellicome J F, Editors) (Cambridge University Press, UK) 43–62

Yang C D, 2000, Design and analysis of composite pipe joints under tensile loading, *J Compos Mater 34* 332–349

Zetterberg T, Astrom B T, Backlund J and Burman M, 2002, On design of joints between composite profiles for bridge deck applications, *Compos Structures 51* 83–91

8.4.8 Manufacturing and fabrication

Astrom B T, 1997, *Manufacturing of Polymer Composites* (Chapman and Hall, London)

Bader M G, Smith W, Isham A B, Rolston J A and Metzner A B, 1990, *Processing and Fabrication Technology: Vol.3 of The Delaware Composites Design Encyclopaedia* (Technomic Publishing Co Inc, Lancaster, Pennsylvania, USA)

Gutowski T W, 1997, *Advanced Composites Manufacturing* (Wiley-Interscience, New York)

Owen M J, Middleton V and Jones I A (Editors), 2000, *Integrated Design and Manufacture using Fibre-Reinforced Polymer Composites* (Woodhead Publishing, Abington, UK)

Spencer R A P, 1983, *Developments in pultrusion, Developments in GRP Technology* (Editor Harris B), (Applied Science Publishers, London) 1–36

Starr T, (Editor), 2000, *Pultrusion for Engineers* (Woodhead Publishing, Abington, UK)

8.4.9 Mechanics

Azzi V D and Tsai S W, 1965, Anisotropic strength of composites, *Proc Society for Experimental Stress Analysis 22* 283–288

Barbero E J and Trovillion J, 1998, Prediction and measurement of the post-critical behaviour of fibre-reinforced composite columns, *Compos Sci and Technol 58* 1335–1341

Barton D C, Soden P D and Gill S S, 1981, Strength and deformation of torispherical ends for GRP pressure vessels, *Int J Pressure Vessels and Piping 9* 285–318

Cox H L, 1952, The elasticity and strength of paper and other fibrous materials, *Brit J Appl Physics 3* 72–79

Lee J, Hollaway L C, Thorne A and Head P 1995, The structural characteristic of a polymer composite cellular box beam in bending, *Construction and Building Mater 9* 333–340

Matthews F L and Davies G A O, 2000, *Finite Element Modelling of Composite Materials and Structures* (Woodhead Publishing, Abington, UK)

Morey T A, Johnson E and Shield C K 1998, A simple beam theory for the buckling of symmetric composite beams including interaction of in-plane stresses, *Compos Sci and Technol 58* 1321–1333

Muc A and Zuchara P, 2000, Buckling and failure analysis of FRP faced sandwich plates, *Compos Structures 48* 145–150

Palmer D W, Bank L C and Gentry T R 1998, Progressive tearing failure of pultruded composite box beams: Experiment and simulation, *Compos Sci and Technol 58* 1353–1359

Sen R, Carpenter W and Snyder D, 1999, Finite-element modelling of fibre-reinforced polymer pre-tensioned elements subjected to environmental loads, *ACI Structural Journal 96* 766–773

Turvey G J, 1997, Analysis of pultruded glass reinforced plastic beams with semi- rigid end connections, *Compos Structures 38* 3–16

Turvey G J, 1999, Flexure of pultruded GRP beams with semi-rigid end connections, *Compos Structures 47* 571–580

Turvey G J, 1999, *Fibre-reinforced polymer composite structural pultrusions and their end connections*, Proceedings of a conference on Composites and Plastics in Construction, Nov 1999, BRE, Watford, UK (RAPRA Technology, Shawbury, Shrewsbury, UK) paper 8 1–8

Turvey G J 1998, Torsion tests on pultruded GRP sheet, *Compos Sci and Technol 58* 1343–1351

Weaver T J, Jensen D W, 2000, Mechanical characterization of a graphite/epoxy IsoTruss, *J Aerospace Engg 13* 23–35

8.4.10   Non-destructive evaluation and smart systems

Culshaw B, 1998, Structural health monitoring of civil engineering structures, *Prog in Struct Engg and Mater 1* 302–315

Kalamkarov A L, Fitzgerald S B, MacDonald D O and Georgiades A V, 1999, On the processing and evaluation of pultruded smart composites, *Compos B: Engg 30B* 753–764

Kalamkarov A L, Fitzgerald S B, MacDonald D O and Georgiades A V, 2000, The mechanical performance of pultruded composite rods with embedded fibre-optic sensors, *Compos Sci and Technol 60* 1161–1170

Kalamkarov A L, MacDonald D O and Fitzgerald S B, 2000, *Performance of pultruded FRP reinforcements with embedded fibre optic sensors*, Proceedings of 4th International Conference on Durability Analysis of Composite Systems (DURACOSYS 99) (AA Balkema, Rotterdam, The Netherlands) 293–301

Lau K T, Zhou L M and Ye L, 1999, Strengthening and strain sensing of rectangular concrete beam using composites and FBG sensors, *Adv Compos Let 8* 323–332

Lau K T, Yuan L, Zhou L M, Wu J and Woo C H, 2001, Strain monitoring in FRP laminates and concrete beams using FBG sensors, *Compos Structures 51* 9–20

Lin M and Chang F K, 1999, Composite structures with built-in diagnostics, *Materials Today 2* 18–22

Sankar J, Hui D, Narayan J, Johnson R, Sibley W and Pasto A, 1999, *Compos B: Engg 30B (7)* 631–773 Interdisciplinary Approach to Smart Composite Structures and Materials (Journal Special Issue)

Schultz M J, Pai P F and Inman D J, 1999, Health monitoring and active control of composite structures by the use of piezo-ceramic patches, *Compos B: Engg, 30B* 713–726

Shi Z Q and Chung D D L 1999, Carbon fibre-reinforced concrete for traffic monitoring and weighing in motion, *Cement and Concrete Research 29* 435–439

Yang S, Gu L and Gibson R F, 2001, Non-destructive detection of weak joints in adhesively bonded composite structures, *Compos Structures 51* 63–71

8.4.11 Mechanical and physical properties and mechanical behaviour

Aggarwal B D and Broutman L J, 1990, Analysis and Performance of Fibre Composites (Second Edition) (John Wiley Interscience, New York)

Amaniampong G and Burgoyne C J, 1994, Statistical variability in the strength and failure strain of aramid and polyester yarns, *J Mater Sci 29* 5141–5152

Brinson H F, 1999, Matrix-dominated time-dependent failure predictions in polymer-matrix composites *Compos Structures 47* 445–456

Bruller O S, 1996, Creep and failure of fabric-reinforced thermoplastics, in *Progress in Durability Analysis of Composite Systems* (Editors Cardon A H, Fukuda H, and Reifsnider K) (AA Balkema, Rotterdam, The Netherlands) 39–44

Curtis P T (Editor), 1988, *CRAG Test Methods for the Measurement of Engineering Properties of Fibre-Reinforced Plastics Composites* (Tech Report 88012) (Ministry of Defence Procurement Executive, Farnborough, UK)

Gassan J and Bledzki A K, 1998, *Possibilities to improve the properties of natural-fibre-reinforced plastics by fibre modification*, Proceedings of the 8th European Conference on Composite Materials ECCM8, Naples, June 1998, Symposium 3, Volume 2 (Woodhead Publishing, Abington, UK) 111–118

Harris B, 1977, Fatigue and accumulation of damage in reinforced plastics, *Composites 8* 214–220

Hashem Z A and Yuan R L, 2000, Experimental and analytical investigations on short GRP composite compression members, *Compos B: Engg 31* 611–618

Kaci S and Toutou Z, 1998, *Prediction of the duration of the relaxation behaviour of a composite cable (Kevlar–49) in relation to temperature*, Proceedings of the 8th European Conference on Composite Materials ECCM8, Naples, June 1998, Symposium 3, Volume 2 (Woodhead Publishing, Abington, UK) 391–396

Moy S S J and Shenoi R A, 1996, Strength and stiffness of FRP Plates, *Proc Inst Civ Eng: Structures and Buildings 116* 204–220

Petras A and Sutcliffe M P F, 1999, Indentation resistance of sandwich beams, *Compos Structures 46* 413–424

Reid S R and Zhou G, (Editors), 2000, *Impact Behaviour of Fibre-Reinforced Composite Materials and Structures* (Woodhead Publishing, Abington, UK)

Soden P D, Leadbetter D and Griggs P R, Eckold G C, 1978, The strength of a filament-wound composite under biaxial loading, *Composites 9* 247–250

Thomson J M, 1982, Electrical properties of carbon-fibre composites, *Practical Considerations of Design Fabrication and Tests for Composite Materials – AGARD Lecture Series no 124* (Director B Harris) (AGARD/NATO Neuilly Paris) paper 9

Zureick A, 1998, FRP pultruded structural shapes, *Prog in Struct Engg and Mater 1* 143–149

8.4.12 Repair and rehabilitation

Andrä H P and Maier M, 2000, *Post-strengthening with externally-bonded pre-stressed CFRP strips*, Proceedings of 16th Congress of the International Association of Bridge and Structural Engineers, Lucerne, Switzerland (IABSE, Zurich) paper 67 382–383

Ascione L, Feo L and Fraternali F, 1998, *Stress analysis of reinforced concrete beams wrapped with FRP plates,* Proceedings of the 8th European Conference on Composite Materials ECCM8, Naples, June 1998, Symposium 3, Volume 2 (Woodhead Publishing, Abington, UK) 197–204

Ascione L and Feo L, 2000, Modelling of composite/concrete interface of RC beams strengthened with composite laminates, *Compos B: Engg 31* 535–540

Black S, 2000, Contractors take composites to new lengths, *Compos Technol 6 (5)* 26–30

Blaschko M and Zilch K, 1998, *Rehabilitation of corrosion-damaged columns with CFRP*, Proceedings of the 8th European Conference on Composite Materials ECCM8, Naples, June 1998, Symposium 3, Volume 2 (Woodhead Publishing, Abington, UK) 371–376

Darby J, Taylor M, Luke S and Skwarski A, 1999, *Stressed and unstressed advanced composite plates for the repair and strengthening of structures*, Proceedings of a conference on Composites and Plastics in Construction, Nov 1999, BRE, Watford, UK (RAPRA Technology, Shawbury, Shrewsbury, UK) paper 13 1–5

Farmer N and Shaw M, 1999, *Controlled supply and application of FRP composite reinforcement for external strengthening*, Proceedings of a conference on Composites and Plastics in Construction, Nov 1999, BRE, Watford, UK (RAPRA Technology, Shawbury, Shrewsbury, UK) paper 12

Feo L, d'Agostino G and Tartaglione D, 1998, *On the static behaviour of reinforced concrete beams wrapped with FRP plates: an experimental investigation*, Proceedings of the 8th European Conference on Composite Materials ECCM8, Naples, June 1998, Symposium 3, Volume 2 (Woodhead Publishing, Abington, UK) 189–196

Focacci F, Nanni A, Farina F, Serra P and Canneti C, 1998, *Repair and rehabilitation of an existing RC structure with CFRP sheets*, Proceedings of the 8th European Conference on Composite Materials ECCM8, Naples, June 1998, Symposium 3, Volume 2 (Woodhead Publishing, Abington, UK) 51–58

Fukuyama H and Sugano S, 2000, Japanese seismic rehabilitation of concrete buildings after the Hyogoken-Nanbu Earthquake, *Cement and Concrete Composites 22* 59–79

Fukuyama K, Higashibata Y and Miyauchi Y, 2000, Studies on repair and strengthening methods of damaged reinforced concrete columns, *Cement and Concrete Composites 22* 81–88

Garden H N, Hollaway L C and Thorne A M, 1998, The strengthening and deformation behaviour of reinforced concrete beams upgraded using prestressed composite plates, *Mater and Structures 31* 247–258

Gilstrap J M and Dolan C W 1998, Out-of-plane bending of FRP-reinforced masonry walls, *Compos Sci and Technol 58* 1277–1284

Hastak M and Halpin D W, 1998, *A new approach for evaluating competing options for composites for retro-fit projects*, Proceedings of the 8th European Conference on Composite Materials ECCM8, Naples, June 1998, Symposium 3, Volume 2 (Woodhead Publishing, Abington, UK) 181–188

Hoppel C P R, Bogetti T A and Gillespie J W 1997, Design and analysis of composite wraps for concrete columns, *J Reinf Plastics and Compos 16* 588–602

Hutchinson R, Donald D, Abdelrahman A and Rizkalla S, 1998, *Shear strengthening of pre-stressed concrete bridge girders with bonded CFRP sheets*, Proceedings of the 8th European Conference on Composite Materials ECCM8, Naples, June 1998, Symposium 3, Volume 2 (Woodhead Publishing, Abington, UK) 43–50

Jai J, Springer G S, Kollar L P and Krawinkler H, 2000, Reinforcing masonry walls with composite materials, *J Compos Mater, 34* 1369–1381

Kachlakev D I and McCurry D D, 2000, Behaviour of full-scale reinforced concrete beams retrofitted for shear and flexural with FRP laminates, *Compos B: Engg 31* 445–452

Kaliakin V N, Chajes M J and Januszka T, 1996, Analysis of concrete beams reinforced with externally bonded woven composite fabrics, *Compos B: Engg 27B* 235–244

Karbhari V M and Seible F, 2000, Fibre-reinforced composites – Advanced materials for the renewal of civil infrastructure, *Appl Compos Mate, 7* 95–124

Keble J, 1999, *Alternative structural strengthening with advanced composites*, Proceedings of a conference on Composites and Plastics in Construction, Nov 1999, BRE, Watford, UK (RAPRA Technology, Shawbury, Shrewsbury, UK) paper 18 1–8

La Tegola A and Noviello G, 1998, *Shear behaviour of concrete beams reinforced with FRP wraps*, Proceedings of the 8th European Conference on Composite Materials ECCM8, Naples, June 1998, Symposium 3, Volume 2 (Woodhead Publishing, Abington, UK) 205–210

Lagoda G and Lagoda M, 2000, *Bridge strengthening by reinforcement bonding – strength and aesthetics*, Proceedings of 16th Congress of the International Association of Bridge and Structural Engineers, Lucerne, Switzerland, (IABSE, Zurich), paper 299 (384–385)

Lees J M and Burgoyne C J, 2000, Analysis of concrete beams with partially bonded composite reinforcement, *ACI Structural Journal 97* 252–258

Li V C, Horii K, Kabele P, HKanda T and Lim Y M, 2000, Repair and retrofit with engineered cementitious composites, *Engg Fracture Mechanics 65* 317–334

Liu H K, Tai N H and Chen C C, 2000, Compression strength of concrete columns reinforced by non-adhesive filament-wound hybrid composites, *Compos A: Appl Sci and Manufg, A31* 221–233

Luyckx J, Lacroix R and Fuzier J P, 1998, *Bridge strengthening by carbon fibres*, Proceedings of the 8th European Conference on Composite Materials ECCM8, Naples, June 1998, Symposium 3, Volume 2 (Woodhead Publishing, Abington, UK) 15–20

Machida A, Mutsuyoshi H and Adhikary B B, 2000, *Recent developments in the repair and strengthening of concrete*, Proceedings of 16th Congress of the International Association of Bridge and Structural Engineers, Lucerne, Switzerland (IABSE, Zurich) paper 361 (362–363)

Mangat P S and O'Flaherty F J, 2000, Influence of elastic modulus on stress redistribution and cracking in repair patches, *Cement and Concrete Research 30* 125–136

Meier U, 1987, Bridge repairs with high-performance composite fibre materials, *Material und Technik 15* 225–228

Meier U *et al*, 1993, Strengthening of structures with advanced composites, in *Alternative Materials for the Reinforcement and Pre-Stressing of Concrete* (Clarke J L, Editor) (Blackie Academic and Professional, Glasgow, UK) 153

Meier U, 1994, Rehabilitation and retrofitting of existing structures through external bonding of thin carbon-fibre sheets, in *Bridge Assessment, Management and Design* (Barr B I G, Evans H R and Harding J E, Editors) (Elsevier Science, Oxford) 373–378

Meier U, 1995, Strengthening of structures using carbon-fibre/epoxy composites, *Construction and Building Mater 6* 341–351

Meier U, 2000, Composite materials in bridge repair, *Appl Compos Mater 7* 75–94

Mendes Ferreira A J, Torres Marques A and Cesar de Sa J, 2000, Analysis of reinforced concrete with external composite strengthening, *Compos B: Engg 31* 527–534

Mikami H, Kishi N and Kurihashi Y, 2000, *Flexural bonding properties of FRP sheet adhered to RC beams*, Proceedings of 16th Congress of the International Association of Bridge and Structural Engineers, Lucerne, Switzerland (IABSE, Zurich) paper 252 (384–385)

Mirmiran A and Shahawy M A, 1997, Behaviour of concrete columns confined by fibre composites, *J Struct Engg-ASCE 123* 583–590

Moriarity J and Barnes F 1998, The use of carbon fibre composites in the London Underground Limited Civil Infrastructure Rehabilitation Program, *SAMPE Journal 34* 23–28

Mosallam A S, 1998, *Seismic performance of welded-frame joints retrofitted with polymer composites and adhesively bonded steel stiffeners*, Proceedings of the 8th European Conference on Composite Materials ECCM8, Naples, June 1998, Symposium 3, Volume 2, (Woodhead Publishing, Abington, UK) 309–322

Mosallam AS, 2000, Strength and ductility of reinforced concrete moment frame connections strengthened with quasi-isotropic laminates, *Compos B: Engg 31* 481–497

Mullen C L, Rice J R, Hackett R M and Ma G L 1999, Filament wound retrofitting of reinforced concrete bridge structures, *J Reinf Plastics and Compos 18* 1113–1121

Nanni A and Norris M S 1995, FRP-jacketed concrete under flexure and combined flexure/compression, *Construction and Building Mater 9* 273–281

Nanni A and Gold W, 1998, *Strengthening of PC slabs with CFRP composites*, Proceedings of the 8th European Conference on Composite Materials ECCM8, Naples, June 1998, Symposium 3, Volume 2 (Woodhead Publishing, Abington, UK) 83–88

Neale K W, 2000, FRPs for structural rehabilitation: a survey of recent progress, *Prog in Struct Engg and Mater 2* 133–138

Ogata T and Osada K, 2000, Seismic retrofitting of expressway bridges in Japan, *Cement and Concrete Composites 22* 17–27

Pantelides C P, Marriott N, Gergely J and Reaveley L D, 2000, *Seismic rehabilitation of damaged bridge piers with FRP composites* Proceedings of 16th Congress of the International Association of Bridge and Structural Engineers, Lucerne, Switzerland (IABSE, Zurich) paper 4 372–373

Quantrill R J and Hollaway L C 1998, The flexural rehabilitation of reinforced concrete beams by the use of prestressed advanced composite plates, *Compos Sci and Technol 58* 1259–1275

Rahman M K, Baluch M H and Al-Gadhib A H, 2000, Simulation of shrinkage distress and creep relief in concrete repair, *Compos B: Engg 31* 541–553

Recuero A and Miravete A 1997, Project on strengthening of structures using advanced composites, *Materiales de Construccion 47* 107–110

Restreppo J L, Wang Y C, Irwin R W and DeVino B, 1998, *Fibreglass/epoxy composites for seismic upgrading of reinforced concrete beams with shear and bar curtailment deficiencies*, Proceedings of the 8th European Conference on Composite Materials ECCM8, Naples, June 1998, Symposium 3, Volume 2 (Woodhead Publishing, Abington, UK) 59–68

Schwegler G and Berset T, 2000, *The use of pre-stressed CFRP laminates as post-strengthening*, Proceedings of 16th Congress of the International Association of Bridge and Structural Engineers, Lucerne, Switzerland (IABSE, Zurich) paper 209 384–385

Shahawy M A, Beitelman T, Arockiasamy M and Sowrirajan R, 1996, Experimental investigation on structural repair and strengthening of damaged pre-stressed concrete slabs utilizing externally bonded carbon laminates, *Compos B: Engg 27B* 217–224

Shahawy M A, Beitelman T, Arockiasamy M and Sowrirajan R, 1996, Reinforced concrete rectangular beams strengthened with CFRP laminates, *Compos B: Engg 27B* 225–233

Sherwood E G and Soudki K A, 2000, Rehabilitation of corrosion damaged concrete beams with CFRP laminates-a pilot study, *Compos B: Engg 31* 453–459

Soutis C, Duan D M and Goutas P 1999, Compressive behaviour of CFRP laminates repaired with adhesively bonded external patches, *Compos Structures 45* 289–301

Spadea G, Bencardino F and Swamy R N, 2000, Optimizing the performance characteristics of beams strengthened with bonded CFRP laminates, *Mater and Structures 33* 119–126

Tai N H, Liu H K and Chen Z C, 2000, Compression after impact (CAI) strength of concrete cylinders reinforced by non-adhesive filament wound composites, *Polymer Compos 21* 268–280

Taljsten B and Elfgren L, 2000, Strengthening concrete beams for shear using CFRP materials: evaluation of different application methods, *Compos B: Engg 31* 87–96

Triantafillou T C, 1998, Composites: A new possibility for the shear strengthening of concrete, masonry and wood, *Compos Sci and Technol 58* 1285–1295

Triantafillou T C, 1998, Strengthening structures with advanced FRPs, *Prog in Struct Engg and Mater 1* 126–134

Wu Z S and Yoshizawa H 1999, Analytical/experimental study on composite behaviour in strengthening structures with bonded carbon fibre sheets, *J Reinf Plastics and Compos 18* 1131–1155

Xiao Y and Wu H, 2000, Compressive behaviour of concrete confined by carbon fibre composite jackets, *J Mater in Civ Engg 12* 139–146

Zhang S, Ye L and Mai Y W, 2000, A study of polymer composite strengthening systems for concrete columns, *Appl Compos Mater 7* 125–138